BestMasters

Mit „**BestMasters**" zeichnet Springer die besten Masterarbeiten aus, die an renommierten Hochschulen in Deutschland, Österreich und der Schweiz entstanden sind. Die mit Höchstnote ausgezeichneten Arbeiten wurden durch Gutachter zur Veröffentlichung empfohlen und behandeln aktuelle Themen aus unterschiedlichen Fachgebieten der Naturwissenschaften, Psychologie, Technik und Wirtschaftswissenschaften. Die Reihe wendet sich an Praktiker und Wissenschaftler gleichermaßen und soll insbesondere auch Nachwuchswissenschaftlern Orientierung geben.

Springer awards "**BestMasters**" to the best master's theses which have been completed at renowned Universities in Germany, Austria, and Switzerland. The studies received highest marks and were recommended for publication by supervisors. They address current issues from various fields of research in natural sciences, psychology, technology, and economics. The series addresses practitioners as well as scientists and, in particular, offers guidance for early stage researchers.

Anna Wolfram

Extending the Complexity of the Leaky Pipeline Phenomenon in Natural Science

A Qualitative Study

Anna Wolfram
HTW Berlin Business School
University of Applied Sciences
Berlin, Germany

ISSN 2625-3577 ISSN 2625-3615 (electronic)
BestMasters
ISBN 978-3-658-43085-6 ISBN 978-3-658-43086-3 (eBook)
https://doi.org/10.1007/978-3-658-43086-3

This Springer Gabler imprint is published by the registered company Springer Fachmedien
Wiesbaden GmbH, part of Springer Nature.
The registered company address is: Abraham-Lincoln-Str. 46, 65189 Wiesbaden, Germany

Paper in this product is recyclable.

Acknowledgement

Being a biochemist myself, I wanted to combine the best of both—my background in science and general management. Thanks to my supervisor and first reviewer Prof. Dr. Jürgen Radel, I was able to do so. For that, our fruitful discussions and your critical input, I want to thank you. I truly appreciate the time and support you dedicated towards my project.

I also want to express my gratitude towards my second reviewer Prof. Dr. Kai Reinhard whom I already met in my first semester in 2020. Thanks to your interesting and diverse lectures, I decided to analyse the work environment in natural science / academia, amongst others, with tools I learned in your class. I also want to thank you for putting me in touch with Prof. Radel and thus, building the formal basis of this thesis.

There is a group of 26 amazing female scientists out there, without whom this project would have been impossible. To all the amazing women who dedicated their precious time to my project, thank you so much. It was an honour to talk to all of you.

To my mentor Carlos, amongst all the other things you continuously do for me, there is one thing that stands out after this thesis. Back in 2018, you once said to me: "In the lab, there are no women and men, in here we are all scientists." You have never reduced anyone to their sex. For you, the only things that count are personal motivation, dedication and achievements. You provide a safe lab space where all scientists can thrive. Thank you for that and everything else.

I also want to mention all the amazing female scientist that have enabled me to research and thrive in natural science / academia: Alex—my very first supervisor in an international lab experience, Victoria—my bachelor's thesis supervisor and Lisa—my PhD thesis co-supervisor. Working with you brought and brings me great joy and makes me want to stay in natural science / academia.

Last but not least, to my husband, my daughter and my parents, thank you for giving me the time and space to take on this journey of my postgraduate MBA in general management.

Abstract

Science, technology, engineering and mathematics (STEM) majorly contribute to global innovation, digitalisation and automatization and thus, drive the global shift towards knowledge-based societies and economies. In this context, it is crucial to utilise all available human capital to further ongoing knowledge-driven development and growth. However, to this day, gender disparities persist in STEM; statistical evidence demonstrates significant female underrepresentation in senior positions—a phenomenon metaphorically described as a *leaky pipeline*. Among other, the *goal congruity theory* is commonly adduced to explain female underrepresentation in STEM. In short, women seek roles to fulfil their valued goals. In result to socio-cultural pressures and psychological orientation, women tend to endorse communal goals more than men. Decreased communal goal affordance associated with STEM fields may disproportionately promote women to opt out of STEM. To test the *goal congruity theory* for coherence in natural science / academia (as a lab-intensive part of STEM), the here presented thesis raises two questions: 1) Does the *leaky pipeline phenomenon* continue to exist in natural science / academia and if so, where are major leaks in the pipeline? and 2) Do female scientists leave natural science / academia to afford other goals? For the former, statistics from 2015–21 unveil near gender parity in early career stages in Germany, the EU and USA. However, female shares drastically decline with increasing career levels in natural science. In particular, professorships are predominantly held by men across all evaluated regions revealing male overrepresentation in decision-making positions. For the latter, 26 female scientists are interviewed in a qualitative study to unveil future aspirations, perceptions of the scientific community as well as perceived reasons to either stay in or leave natural science / academia.

Among study participants, an overall positive as well as communal perception of natural science / academia is identified. In addition, the perceived affordance of communal values within science majorly contributes to the desire of female researcher to stay in the academic path. These findings contradict the literature-established *goal congruity theory* and thus, provide an experienced-based foundation to extend the complexity of the *leaky pipeline phenomenon* in natural science.

Preliminary Remarks

During my research project, I had the amazing opportunity to talk to wonderful female scientists from all around the world. All these amazing women hold their own space in the diverse and multi-national scientific community. To get there, they had to face scientific and other challenges that are to some extend part of natural science (e.g., experimental failure, paper and grant rejections, etc.). During the very honest and personal interviews, I greatly profited from their experience and wisdom. To share their expertise with the greater community, I have asked all of them to share a piece of advice for up-and-coming (female) researchers. Below, you can find their answers. May it help many to persevere and thrive in natural science / academia.

Embrace challenges. You are needed in science because you have something to bring to the field. You belong here and we are looking forward to having you.

Everyone at this level is smart. Distinguish yourself by being kind.

Scientific research requires a lot of hard work and dedication for comparatively low pay, but nothing beats the highs of when you have painstakingly designed every experiment to drive the partnering of an anti-cancer agent, and that agent allows a young person of similar age to achieve enough reprieve to walk again when they were only given months to live with no real therapeutic options.

Try to do what you would like to do in in the future.

Build an effective and supportive network as early as possible. Trust yourself. There is no bad decision if it is yours; even if it ends up being a mistake, you will be able to course-correct. A career is not always linear and is often uncertain, so embrace the uncertainty of it and all the surprises that come along the way.

Do what really interests you. If you do what's interesting to you, the time is always well spent.

Know yourself really well so that you can strengthen your strength. Step out in the community alone as an individual and network. Network, network, network, because this is what in the end will help you to really get to the positions that you want.

Don't be afraid to fail.

Believe in yourself and your right to have a place at the table.

Have your own vision of success.

Find colleagues that actually support you. Surround yourself with excellent people—mentors, collaborators, co-workers—who will both support and challenge you with your best interests at heart.

Don't get stuck on what you think is a successful career. Don't miss out on opportunities because you are too focused on the end of the corridor.

Believe in yourself. Promote yourself. Invest in yourself.

Follow your dreams.

As a woman in academia, you have to survive, perseverance is the key. Find people to talk to and you will find a way because there are so many people in the world, your case cannot be unique.

Love what you do and make sure to follow what you love to do. Be aware of your position in the academic world. Support [others] wherever needed.

Maintain a work-life-balance and make sure to have other things in your life other than science.

Talk to people.

Try to keep leering all the time.

Write down why you do what you do. Revisit it every year and decide whether you're still fulfilling that mission.

Don't stop to pursue your goals. Just put your wings on and fly. No one is born to appease others.

Never put science above yourself.

Science should never be your number one. Be careful with passion for science.

Prepare yourself for a difficult ride but one that is extremely rewarding at the end. Pick really good mentors. Drive change in a positive way because if you're silent and you just endure it, nothing changes. Cultivate strong relationships around you. Don't ever underestimate happiness.

Find the right people, find the right advisor and choose the best lab. It's more important to find the right people that will support you in your early career than it is to study the exact right project.

You can't fix the system if you are not in the system. Do what you need to get your letters of reference and get yourself through. Don't burn yourself out trying to change a system or a culture in which you hold no power.

Contents

Abbreviations

B.Sc.	Bachelor of Science
CEOs	Chief Executive Officers
COVID-19	Coronavirus disease
D	Germany
EMS	Electronic Supplementary Material
EU	European Union
EWOB	European Women on Boards
FüPoG	Führungspositionen-Gesetz
ISCED	International Standard Classification of Education
M.Sc.	Master of Science
NGO	Non-governmental organisation
NIH	National Institutes of Health
NSF	National Science Foundation
OECD	Organisation for Economic Co-operation and Development
PhD	Doctor of Philosophy
PI	principal investigator
Postdoc	post-doctoral researcher
Q3	third quarter
R&D	Research and development
STEM	Science, Technology, Engineering and Mathematics
U.S.	United States
UK	United Kingdom
UN	United Nations

List of Figures

Introduction

1

Historical gender roles continue to impact socially accepted perceptions of both men and women. To this day, women are seen as the main caregiver; they are described as warm rather than competent. Amongst others, these gendered stereotypes impede with women's career aspirations. As a consequence, vertical gender segregation—the underrepresentation of women in higher career levels—is seen across various working fields. Metaphorically, this has been described as the "leaky pipeline" phenomenon, which particularly persists in the fields of science, technology, engineering and mathematics (STEM). In the following section, three interrelated hypotheses are introduced to the reader as possible explanations for female underrepresentation in STEM.

1.1   Historical Gender Roles in the 21ˢᵗ Century

In past centuries, men have dominated public and academic life while women were mostly confined to their family lives. Even though these historical gender roles are perceived as antiquated by many, they have shaped social and cultural values that persist to this day.

According to the *International Labour Office* (2016), women spend at least 2.5-fold more time on unpaid household and care work than men. This value presents the statistical average for 23 developing and 23 developed countries and thus, underlines that caregiving is still perceived as a female role. In fact, gendered stereotypes persist to the extent that scientists have identified a *motherhood penalty* and a *fatherhood bonus* for women and men with children. Statistically, women lose approx. 5% of wage for every child even when accounting for additional factors (e.g., work hours and care work) (Benard et al., 2007; Correll et al., 2007). When researchers submitted identical applications only differing in parental status, mothers were offered 7% less than childless women. This accounts to a gross difference of approx. $ 11,000 per year. More strikingly, male applicants with children were offered approx. $ 13,000 (9%) more annual income than women with children (Correll et al., 2007). These pay gaps demonstrate the societal perception that men need to financially provide for the family while women must care for the children and thus, are less able to work committedly (Correll et al., 2007; Cuddy et al., 2004).

In 2004, Cuddy *et al.* performed a *study of stereotypes of working moms in professional settings* with 122 Princeton University undergraduates. The results demonstrated that even young people believe that female professionals trade competence for warmth when having children and are therefore perceived to be less hireable, less promotable and less worthy of additional training relative to working dads and workers without children (Cuddy et al., 2004).

1.2   Vertical Gender Segregation in Management

The phenomenon of female underrepresentation in top levels of occupational hierarchies despite the academic qualifications is known as vertical gender segregation. Based on the socially accepted gender roles / stereotypes described above, it seems little surprising that vertical gender segregation persists in management across various organisations and nations (Seo et al., 2017). Across EU member states, women held 34% of management positions in the third quarter (Q3) of 2020. Germany—a worldwide leading economy—even undercuts the EU average by 3 percentage points. The respective female quotes of all EU countries are depicted in **Figure 1.1** (eurostat, 2021). One year later, European Women on Boards (EWOB) analysed 668 companies (2021). On average, women accounted for 19% of executive level of company decision-makers. More strikingly, only 50 Chief Executive Officers (CEOs) were women amounting to a female share of 7.5% (European Women on Boards, 2021).

In fact, female underrepresentation is a well-recognised and researched phenomenon that has extensively been studied and reviewed over the last years (Auschra et al., 2022; Ayub et al., 2019; Babic & Hansez, 2021; Fernandez-Mateo & Fernandez, 2016; Oakley, 2000; Seo et al., 2017). In addition, data and reports published by intergovernmental organisations (e.g., *The World's Women 2020: Trends and Statistics* by United Nations (UN), OECD *Gender Data Portal* by Organisation for Economic Co-operation and Development (OECD)) and big management consultancies (e.g., *Women in the Workplace 2021* by McKinsey) regularly raise public awareness on female underrepresentation in management.

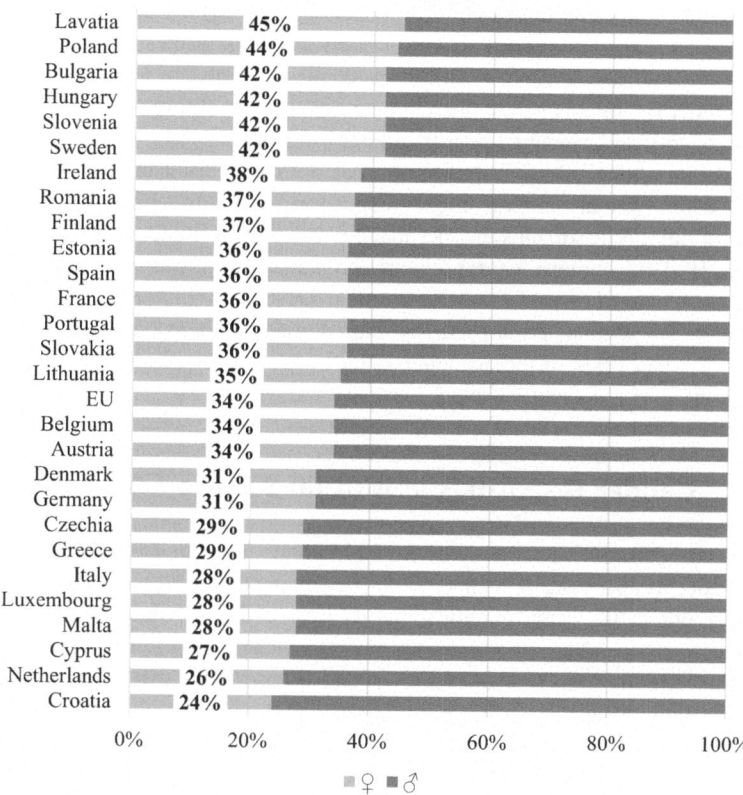

Figure 1.1 Gender distribution in management across EU countries. (Reference: (eurostat, 2021). Author's own graph)

1.3   Female Quota to Counteract Vertical Gender Segregation

Amongst other, activism, public pressure and broad media coverage have prompted the German government to pass the *German Act on Equal Participation of Women and Men in Leadership Positions in the Private and Public Sectors* (also known as *Führungspositionen-Gesetz I* (FüPoG I)) in 2015. Consequently, 30% of supervisory board representatives of publicly listed companies must be women. In 2021, the Act was updated to enforce a female share of 1 in 3 executive members (FüPoG II). In conclusion, the German government enforces a female quota for supervisory and executive boards in large companies.

However, vertical gender segregation does not only persist in the private and public sectors but also in the higher education sector. In 2021, women held 27.2% of all professorships in Germany (Statistisches Bundesamt (Destatis), 2022, p. 18–19). Amongst natural science and mathematics, women accounted for only 21.5% of professors in the same year (Statistisches Bundesamt (Destatis), 2022, p. 45 f.). To this end, it seems of value to take a closer look at the prevailing gender disparities in the academic career path within natural science (from now on referred to as natural science / academia).

1.4   Gender Disparities in Natural Science / Academia

At the end of the 19th, beginning of the 20th century, women were able to access higher education for the first time. From that point onwards, they were able to officially obtain the same qualifications as men. Early on, outstanding female scientists including Marie Salomea Skłodowska—Curie (Marie Curie), Rosalind Franklin, Katherine Johnson and Jane Goodall demonstrated that women, in contrast to common beliefs, made ground-breaking discoveries and thereby drove scientific innovation. Nevertheless, science continued and continues to be perceived as a male domain (Makarova et al., 2019; Makarova & Herzog, 2015; McKinnon & O'Connell, 2020; Reuben et al., 2014). McKinnon *et al.* postulated that prevailing gender biases and stereotypes impede with the attraction, retention and progression of girls and women in STEM (2020). Thus, whilst women have made gains in the STEM fields over the last decades, gender equality has not been reached (Charlesworth & Banaji, 2019; Huang et al., 2020; Tannenbaum et al., 2019). In addition, a *leaky pipeline phenomenon* can be observed (Blickenstaff, 2006). This metaphorical reference describes the loss of women along the academic path towards more senior positions (Goulden et al., 2011; Howe-Walsh &

Turnbull, 2014; Resmini, 2016; Ysseldyk et al., 2019). Today, three interrelated hypotheses are commonly adduced as possible explanations for this phenomenon in natural science / academia (Charlesworth & Banaji, 2019);

1) Innate and / or socially attributed STEM performance differences (Hyde, 2014, 2016; Hyde et al., 2008; Hyde & Mertz, 2009),
2) Innate and / or socially attributed variation in STEM engagement and attitude to life (Ceci et al., 2009, 2014; Ceci & Williams, 2011; Diekman et al., 2010, 2020),
3) Explicit and implicit biases in both genders (Kurdi et al., 2019; Nosek & Smyth, 2007, 2011; Régner et al., 2019).

Below, these proposed hypotheses shall be elaborated in detail.

1.4.1 Innate and / or Socially Attributed STEM Performance Differences

Despite major success of female researchers, the idea that women are less capable scientists remained in the society. Amongst other reasons, female underrepresentation in natural science / academia was and still is explained by innate and / or socially attributed STEM performance differences (Hyde, 2014, 2016; Hyde et al., 2008; Hyde & Mertz, 2009; Keller & Scharff-Goldhaber, 1998). To validate this, researchers tried to demonstrate male superiority in skills perceived as relevant for natural science. In particular, math performance has commonly been adduced to reinforce fundamental difference in men and women (Hyde, 2014). However, the analysis of math abilities of more than 7 million students has identified insignificant differences; mathematical capabilities of both sexes align almost identically (Hyde et al., 2008; Hyde & Mertz, 2009). Put simple, gender-dependent performance differences in math do not exist. In addition, individual math abilities hold little explanatory power to predict the respective success in natural science / academia. In recent years, other relevant STEM skills including spatial and language ability have been evaluated to identify gender difference. In 2007, Halpern et al. postulated that scientists need to effectively communicate their scientific achievements in written and oral form. Here, they speculated that women would have an advantage due to their enhanced writing skills (Halpern et al., 2007). Data from the *U.S. National Assessment of Educational Progress* (based on 3.9 million students) revealed statistically significant advantages for reading (on third of a standard deviation by Grade 12) and writing (on half of a standard deviation by

Grade 12) (Reilly et al., 2019). Whether observed differences are attributed to innate differences (Halpern et al., 2011) or sex-typing of abilities (Marinak & Gambrell, 2010) remains elusive. Respective results on gender-dependent spatial abilities deviate based on subject age, testing format, and test framing (Huguet & Régner, 2009; Voyer, 2011; Voyer et al., 1995, 2017). In conclusion, due to the lack of genuine proof, innate and / or socially attributed STEM performance differences provide an inconclusive explanation for observed gender disparities in natural science / academia.

1.4.2 Innate and / or Socially Attributed Variation in STEM Engagement and Attitude to Life

Important male scientists such as Alfred Nobel, Albert Einstein, Louis Pasteur and Stephen Hawking have shaped the socially accepted idea that scientists are brilliant geniuses who work by themselves to drive scientific progress. In fact, scientists are stereotypically described as intelligent, highly trained and devoted. At the same time, they are narrated as brainy, dull and working alone (Beardslee & O'Dowd, 1961; Schinske et al., 2015). In line with these stereotypes, Diekman *et al.* argue that STEM fields are perceived as competitive rather than collaborative work environments (2010, 2011, 2017, 2020). In turn, this fosters the perception that communal goals such as working with and / or helping others cannot be reached. Subsequently, communal goal affordance in STEM is perceived lower than in other working fields (Brown et al., 2015). Diekman *et al.* further postulate that humans seek roles which they believe to be in line with their valued goals—a concept introduced as *goal congruity hypothesis* (Diekman et al., 2010, 2011) and later referred to *goal congruity theory* or *goal congruity perspective* (Diekman et al., 2020). A long list of research is quoted in order to underline this theory (Brown et al., 2015; Diekman et al., 2010, 2011, 2020; Weisgram et al., 2011; Weisgram & Bigler, 2006). Here, it is argued that values are rooted in the early childhood where both boys and girls experience social pressures (some even refer to innate inclinations) to fulfil different roles in society. In result to socio-cultural pressures and psychological orientation, women tend to endorse communal goals more than men. Thus, the above described perception of STEM fields may disproportionately promote women to opt out of natural science / academia (Brown et al., 2015; Diekman et al., 2010, 2017, 2020; Fuesting & Diekman, 2017). In other words, women choose to leave natural science to fulfil goals that they perceive to be mismatched with the nature of academia. Hence, the *goal congruity perspective* can be understood as a theoretical framework to interpret the interdependence of personal choices and social roles.

It is noteworthy that Diekman *et al.* concluded that the activation of communal goals decreased STEM interest in study participants (Diekman et al., 2011). Put simply, they deduced a inverse correlation between the interest in STEM and communal goals.

If true, this would implicate that people pursuing a career in natural science / academia (as part of STEM) are less interested in communal goals. From a female perspective, the *goal congruity perspective* indirectly implies that women who leave science are more interested in communal and women who stay in individual goals. However, the respective study was based on 64 psychology students (37 of them were women) with a median age of 19 years. Thus, it seems questionable to what extend study findings are transferable to scientists working in STEM. In conclusion, further research efforts may be required to demonstrate validity of the *goal congruity perspective* for female scientists.

1.4.3 Explicit and Implicit Biases in Both Genders

Next to difference in performance and attitude, bias is named as a main contributor to prevailing gender disparities in STEM. In particular, biases in funding, authorships and prestigious awards are regularly cited (Kurdi et al., 2019; Nosek & Smyth, 2007, 2011; Régner et al., 2019) and shall be highlighted hereafter.

Funding: When evaluating the gender distribution of grant recipients across major U.S. funding agencies (e.g., National Science Foundation (NSF) and the National Institutes of Health (NIH)), gender parity has been reached (Hosek et al., 2005; Pohlhaus et al., 2011). However, subtle differences persist. For instance, men reapplied / renewed their funding at NSF and NIH more often with 5% and 20% higher reapplication / renewal rates, respectively (Hosek et al., 2005). In addition, female application rates were lower for the top 1% of NIH grants (Hosek et al., 2005). Subsequently, women held smaller grants on average in the past (Hosek et al., 2005; Oliveira et al., 2019; Waisbren et al., 2008). Based on decreased female shares in top grant application and grant renewal, it could be hypothesised that women underpromote or undervalue themselves and their science. In fact, studies have shown that women in STEM are less confident about their own abilities in contrast to men's abilities despite having near-identical human capital (Correll, 2016; Sterling et al., 2020). This phenomenon is also known as *confidence gap* and is largely based on implicit bias within women about their own capabilities as well as those of others.

Authorships: When investigating gender distributions across authorships, some fields have reached gender party (Brooks & della Sala, 2009) and others are on their way (Holman et al., 2018). In other research fields, women stayed underrepresented in comparison to their male colleagues (Holman et al., 2018; Ross et al., 2022; West et al., 2013). In addition, women are underrepresented in last authorships usually representing the group leader or principal investigator (PI). The observed *productivity gap* between men and women may arise due to differences in productivity or lack of acknowledgement. In a recent study, investigators provided evidence that women had to contribute more than men to be included in the author's string. Overall, data indicated that women are less likely to be credited with authorship (Ross et al., 2022).

Prestigious awards: The Nobel Prize in medicine, in chemistry and in physics may be considered as the most prestigious awards in STEM. Here, gross discrepancies between honouring scientific achievements of men and women are detectable. In total, only 12% Nobel laureates in medicine, 3% in chemistry and 1% in physics were women (Charlesworth & Banaji, 2019). When analysing 141 highly prestigious international prizes (e.g., Nobel prizes, the Fields Medal for mathematics and the Robert Koch Award for biomedical sciences) between 2001 and 2020, a total of 2,273 scientists received an award. Out of them, 11.5% were women (262 / 2,273) (Meho, 2021). Thus, to this day, female scientists are less likely to win prestigious awards (Watson, 2021).

In summary, whilst near-gender parity has been reached in funding, positive bias towards men persists in authorships and prestigious awards. In addition, as discussed in **Section 1.1**, biases in hiring, compensation and perceived competence also favour men in STEM. It should be mentioned that intersectional inequalities also persist in STEM (Kozlowski et al., 2022), however, the elaboration would go beyond the scope of this thesis.

1.4.4 Key Takeaway: *Leaky Pipeline Phenomenon* and Proposed Explanations

To this day, the following three hypotheses are commonly consulted to explain the *leaky pipeline phenomenon* in STEM:

A) Innate and / or socially attributed STEM performance differences,
B) Innate and / or socially attributed variation in STEM engagement and attitude to life,
C) Explicit and implicit biases in both genders.

However, there is no conclusive evidence that proves innate and / or socially attributed STEM performance differences in men and women. In addition, studies underlining innate and / or socially attributed variation in STEM engagement and attitude to life were not conducted with female scientist who have actively worked in STEM. Thus, it seems questionable if respective findings relate to women working in STEM. In contrast, clear evidence underlines the persistence of gender differences in STEM. Amongst others, the *pay gap, confidence gap* and *productivity gap* have all been associated to implicit and explicit bias in men and women towards themselves and others.

*In **Chapter 1**, the interrelationship of historical gender roles and modern gender stereotypes was displayed. To this day, women suffer a perceived lack of competence despite near-identical human capital. Amongst others, this contributes to vertical gender segregation in all working fields including STEM. The loss of women in higher position—also referred to as leaky pipeline phenomenon—has commonly been attributed to three interrelated hypotheses. In summary, the presented research unveiled that, 1) due to the lack of coherent evidence, innate and / or socially attributed STEM performance differences between men and women cannot be proven. 2) It is uncertain whether findings, implicating perceived communal goal affordance as a main reason to leave STEM (goal congruity theory), may actually translate to women working in STEM. 3) Both implicit and explicit bias towards men and women reside in both women and men. In a nutshell, based on these findings, the three proposed hypotheses do not present a coherent explanation for the leaky pipeline phenomenon in STEM. Subsequently, the following **Chapter 2** will underline the relevance to challenge the dominating hypotheses.*

Research Focus and Relevance

2

*Gender differences persist in STEM fields. However, the literature-cited explanations discussed in the section above may not actually cover the complexity of the leaky pipeline phenomenon. In **Chapter 2**, the relevance of gender equality to progress in our knowledge-driven societies and economies is demonstrated. To achieve gender parity across all career levels in STEM, educated efforts need to be undertaken. Yet, incomplete knowledge prevents the development of sensible measures to counteract vertical gender segregation. To this end, two hypotheses will be postulated; the first to describe the current state of the leaky pipeline (**Hypothesis 1**) and second to characterise the validity of the goal congruity theory for women working in natural science / academia (**Hypothesis 2**).*

In pursuit of gender equality, educated efforts need to be undertaken to foster gender parity in historically male-dominated working fields. Here, the fields of science, technology, engineering and mathematics deserve particular attention as they majorly contribute to global innovation, digitalisation and automatization and thus, drive the global shift towards knowledge-based societies and economies. In this context, it is crucial to utilise all available human capital to further ongoing knowledge-driven development and growth. As elaborated above, both men and women have near-identical human capital. At the same time, individuals are equipped with unique sets of skills and experiences. Thus, intellectual benefits may be unlocked when reaching equal participation of all qualified humans (Charlesworth & Banaji, 2019). In fact, several studies indicate that scientific output is higher and more impactful for gender diverse teams (Gewin, 2018; Hodapp & Brown, 2018; Prager, 2021; Yang et al., 2022). Hence, it should be a common goal to reach and maintain gender parity across all career levels in STEM (as well as all other aspects of life).

To tackle persisting gender disparities, a comprehensive characterisation of the current state of the *leaky pipeline* as well as underlying causes of female underrepresentation is required. For the latter, the three literature-cited hypotheses commonly adduced to explain the *leaky pipeline phenomenon* in STEM may hold limited validity (see **Section 1.4.4**). The lack of coherent evidence defeats the concept of innate and / or socially attributed STEM performance differences (see **Section 1.4.1**). Whilst innate and / or socially attributed variation in STEM engagement and attitude to life can be proven for men and women in general (see **Section 1.4.2**), it seems somewhat questionable whether these findings translate to women working in science. Among others, *communal goal affordance* (i.e., the ability to help, care for or work with others) is stated as a major reason why women leave science. The *goal congruity theory* defines this as follows; *"individuals seek [...] roles that fulfil their valued goals"* (Diekman et al., 2020). Put simply, women enter the scientific career pipeline, continue to work in the field and finally leave because they feel that they cannot pursue their (mostly communal) goals in the STEM work environment. In short, women leave science due to a perceived mismatch between individual and STEM goals.

If this line of argument was true, one could expect to see women opting out of STEM in early stages or the career. In particular, natural sciences are rather lab intensive. Already bachelor's and master's students spend large fractions for their higher education in laboratories and thus, first-hand experience the nature of natural science / academia. In these fields, it is also highly common to do a research-based PhD. In total, students spend at least 8 years in their field before they can transition to postdoctoral employment. During a timespan of nearly 10 years, people should know whether their individual goals align with their work environment. Thus, if the *goal congruity theory* was true, one should see a decline in female shares along the way of higher education as well as at the transition to postdoctoral employment. In other words, one would not expect constant gender distributions across various career levels. To test the *goal congruity theory*, this thesis will address two main questions:

A) Does the *leaky pipeline phenomenon* continue to exist in natural science / academia and if so, where are major leaks in the pipeline?
B) Do female scientists leave natural science / academia to afford other goals?

In line with these questions, the author of this thesis has postulated two hypotheses which are elaborated below.

2.1   Hypothesis 1: Female Underrepresentation Predominately Persists in Senior Positions in Natural Science / Academia

Female underrepresentation predominately persists in senior positions in natural science / academia. In fact, (near) gender parity is achieved in early career stages. Along the secondary and tertiary education, female shares are approximately constant. Drastic changes in gender distributions are predominately seen in more senior positions. In most senior positions, women are significantly underrepresented.

To probe **Hypothesis 1**, statistical data sets from German as well as international agencies and institutes will be analysed. The focus is set on three global clusters: 1) Germany, 2) the member countries of the European Union (EU) and 3) the United Sates (U.S.) of America. The respective method is described in **Section 3.2.1**.

2.2   Hypothesis 2: The *Goal Congruity Theory* Provides an Incomplete Explanation for Female Underrepresentation in Natural Science

The *goal congruity theory* provides an incomplete explanation for female underrepresentation in natural science. In fact, a majority of female scientists does not leave natural science / academia because of a perceived mismatch between individual goals and the nature of academia. Rather, the ongoing lack of funding and lack of positions create an instable work environment, in which many scientists cannot sustain, let alone progress.

To qualitatively test **Hypothesis 2**, an international qualitative study will be designed and conducted. Firstly, the study will try to unveil future aspirations of female scientists that are currently employed as postdoctoral researchers (postdocs) or junior (research) group leaders. Please note that these positions were chosen as they represent temporary positions at which women (as well as men) have to decide to either stay in or leave natural science / academia. Secondly, a focus will be set on women's perceptions of the scientific community. Thirdly, perceived reasons to either stay in or leave natural science / academia shall be characterised. Afterwards women will be confronted with a statement that represent the *goal congruity theory*:

Women choose to leave natural science to fulfil goals that they perceive to be mismatched with the nature of academia.

In the end, individual perceptions and responses to the *goal congruity theory* shall be carefully contrasted. The respective design and conduct of the qualitative study are described in **Section 3.2.2**.

*The goal congruity theory offers an explanation for female underrepresentation in STEM; Women seek roles to fulfil their valued goals. Here, women over-proportionally endorse communal goals they perceive to be unaffordable in STEM. Consequently, women are more likely to leave their role (the role being a scientist in STEM). To test the goal congruity theory for coherence in natural science / academia (as a lab-intensive part of STEM), the here presented thesis aims to address two major questions: 1) Does the leaky pipeline phenomenon continue to exist in natural science / academia and if so, where are major leaks in the pipeline? and 2) Do female scientists leave natural science / academia to afford other goals? Two respective hypotheses have been postulated by the author (**Hypothesis 1** and **Hypothesis 2**). In the subsequent **Chapter 3**, the respective material and methods required will be highlighted.*

Materials and Methods

3

*In the following **Section 3**, all utilised materials (**Section 3.1**) and methods (**Section 3.2**) are described. The reader gains an understanding of used techniques to achieve the research aims discussed above. To allow for easy readability, methods are allocated to either **Hypothesis 1** or **Hypothesis 2**. In short, statistical data analysis of national and international agencies and institutes as well as the setup of an international qualitative study are elaborated. For the latter, study design, setup and conduction are highlighted. A special focus is set on the selection and invitation of eligible study participants. Taken together, **Section 3** provides the comprehensive methodologic foundation of the here presented master's project.*

3.1 Materials

To complete this research project and corresponding thesis, the following software (**Section 3.1.1**) and transcription service (**Section 3.1.2**) were used.

3.1.1 Software

Table 3.1 lists the computer software used in the here presented research project.

Supplementary Information The online version contains supplementary material available at https://doi.org/10.1007/978-3-658-43086-3_3.

A. Wolfram, *Extending the Complexity of the Leaky Pipeline Phenomenon in Natural Science*, BestMasters, https://doi.org/10.1007/978-3-658-43086-3_3

Table 3.1 Software

Software	Version
Microsoft Teams	1.5.00.22362
Zoom	5.11.6
Voice Memos	2.3
Microsoft Word	16.65
Microsoft Excel	16.65
Mendeley Reference Manager	2.79.0
ATLAS.ti	22.1.0

3.1.2   Transcription Service

Rev Transcription Services were used to automatically transcribe all audio files into ready-to-use text files (90+ % accuracy). Afterwards, audio files and transcripts were manually cross-checked by the study investigator.

3.2   Methods

3.2.1   Hypothesis 1: Statistical Data Analysis

To address **Hypothesis 1**, female representation in natural science / academia was analysed in Germany (D), the European Union (EU) and the United States (U.S.). To this end, reports from German (Krabel et al., 2021), European (European Commission & Innovation Directorate-General for Research, 2021) and U.S. agencies and institutes (Fry et al., 2021; Office of Federal Operations, 2019) were evaluated. Based on the respective statistical data sets, female shares across typical career levels in natural science / academia were deduced and plotted using *Microsoft Excel*. To ensure facile comparability, the chart type *100 % stacked bar chart* was chosen as the standard chart for all data corresponding to **Hypothesis 1**. In addition, selected data sets were also plotted using a *line with markers chart*. Based on data availability and country-dependent specifics, gender distributions were included for the following career levels.

- **General:**

    o Female and male share of Bachelor of Science (B.Sc.) recipients,
    o Female and male share of Master of Science (M.Sc.) recipients,

○ Female and male share of doctoral students (PhD students),

○ Female and male share of doctorate recipients (PhD),

○ Female and male share in STEM jobs,

○ Female and male share of STEM / natural science faculty staff,

- **Germany:**

  ○ Female and male share of Habilitation recipients,

  ○ Female and male share of newly appointed Junior professorships,

  ○ Female and male share of newly appointed W2 professorships,

  ○ Female and male share of newly appointed W3 professorships.

- **U.S.**

  ○ Female and male share of Lecturers/ Senior Lecturers,

  ○ Female and male share of Assistant Professors,

  ○ Female and male share of Associate Professors,

  ○ Female and male share of Professors.

- **European Union**

  ○ Female and male share of ISCED 6 and 7 students (6: B.Sc. or equivalent; 7: M.Sc. or equivalent),

  ○ Female and male share of ISCED 6 and 7 graduates (6: B.Sc. or equivalent; 7: M.Sc. or equivalent),

  ○ Female and male share of ISCED 8 students (PhD student or equivalent),

  ○ Female and male share of ISCED 8 graduates (PhD or equivalent),

  ○ Female and male share of Grade C staff (first position of a newly qualified PhD),

  ○ Female and male share of Grade B staff (research staff not as senior as the top position (A) but more senior than newly qualified PhDs (C)),

  ○ Female and male share of Grade A staff (single highest position at which research is normally conducted within the institutional or corporate system).

When applicable, primary and secondary data sources were stated below each chart. The obtained data sets depict gender distribution between 2015–2021. The contribution of other factors such as age, marital status, ethnic background, socio-economical background or intersectionality were not scope of this thesis. The respective data search was performed in September 2022. Based on reference-dependent nomenclature, respective female shares are shown for *typical academic career* (term from *She figures 2021:gender in research and innovation:statistics and indicators*, not clearly defined), **STEM** (science, technology, engineering and mathematics), **natural science** (chemistry, astronomy, earth science, physics, and biology), **natural science and mathematics**, **life sciences** (scientific studies of life, e.g., biology) and **physical sciences** (chemistry, astronomy, earth science and physics) in **Section** 4.1.

3.2.2   Hypothesis 2: Qualitative Study

To address **Hypothesis 2**, an international qualitative study was designed to char-acterise aspects of early-stage researchers' personal experiences related to career choice and career advancement in natural science / academia. To obtain com-prehensive information on individual experiences, perceptions and reflections in addition to basic demographic data of study participants, a problem-centred inter-view was chosen—a qualitive interview methodology first introduced by Andreas Witzel (1982, p. 66 ff.) and later described by Cornelia Helfferich (2011, p. 13–107). Early-stage researchers were defined as postdoctoral researchers (postdocs) and junior (research) group leaders or equivalent (e.g., core facility staff, lec-turer) in natural science / academia. In addition, former postdocs that had recently (within the last 3 years) transitioned to industry were also considered as eligible to participate.

3.2.2.1   Recruitment of Study Participants

Study participants were recruited via direct email invitation (see **EMS**, p. 2). To minimise region-dependent bias, women from all around the world were invited. To access a large number of contact details, the author pursued three strategies:

A) Contacting postdocs/ junior group leaders within the scientific network of the author (5 / 68 invited women, 4 / 26 study participants),
B) Contacting senior staff within the scientific network of the author to access their network and contacts (15 / 68 invited women, 8 / 26 study participants),
C) Searching *Gage*—the world's largest directory of women and gender diverse folks in science (Gage. Discover Brilliance)—for female scientists from around the globe to invite them and access their network and contacts (48 / 68 invited women, 14 / 26 study participants).

In total, 68 women were invited to participate. 26 / 68 agreed to participate. The participants were briefed about the interview conduction and subsequent data analysis and handling in written text. Before the date of the interview, all signed a personalised interview consent form (see **EMS**, p. 4).

3.2.2.2 Interview Guide, Setup and Conduction

Based on an approach previously reported by Cornelia Helfferich (2011, S. 178–186), a semi-structured pilot interview guide was designed (see **EMS**, p. 6). The interview guide was divided in two parts: I) problem-centred data and II) demographic data.

The initial interview guide for part I contained four leading questions:

A) *Could you please tell me why you personally chose to pursue a career as a scientist?*
B) *As a Postdoc/junior group leader you hold a great track record in academia. Would you please share your future aspirations with me?*
C) *How would you describe the scientific community? Could you please name 3 to 5 attributes and briefly elaborate them?*
D) *Would you please comment on the following statement?*
   Women choose to leave natural science to fulfil goals that they perceive to be mismatched with the nature of academia.

The questions were designed rather broadly to invite participants to freely share their experiences. Participants did not receive the questions prior to the interview to ensure impromptu responses. All interviews were set up in a semi-formal atmosphere as one-on-one video calls via *Microsoft Teams* or *Zoom*. The study was conducted in two phases: a 4-interview pilot phase followed by interview guide enhancement and another 22 interviews. The pilot phase was conducted between the 20th–30th of May 2022.

Based on participants' responses, the pilot interview guide (Part I) was modified to improve the transition between leading question 3 and 4. To this end, the following question was included in between question 3 and 4:

Many scientists have to face the choice to either stay in academia / or leave academia. From a personal perspective, could you name relevant factors that may affect this choice?

The second interview phase was conducted between June and mid-September 2022. All 26 interviews were audio recoded using *Voice Memos*.

3.2.2.3 Interview Transcription and Analysis

All audio recordings were automatically transcribed by *Rev Transcription Services* to obtain ready-to-use text files (90+ % accuracy). Then, corresponding transcripts and original audio files were cross-checked for accuracy by the investigator. All content errors were corrected. Revised transcripts were exported as *Word* files. Using a self-designed coding system, relevant information (see **EMS**, p. 9, "**Transcript analysis: Collected data for statistical analysis**") were extracted from each transcript and unified in one table. Afterwards, all study participants received an email with their audio recording, transcript and table (see **EMS**, p. 9) with their personal information. All of them had the chance to revise all files and report or correct any information missing or invalid. Out of 26, 11 corrected and / or confirmed their data sets before October 2022. Afterwards, an updated comprehensive *Excel* file was generated as a data basis for further statistical analysis. To protect study participants from discrimination or other harm, all data sets were anonymised. Primary data were pooled and statistically analysed to show global trends rather than individual findings (see **Section** 4.2).

*In **Section 3**, both, the national and international data report analysis (**Hypothesis 1**) and the set-up of an international qualitive study (**Hypothesis 2**) were elaborated. In the former, German, European and United States reports were evaluated to obtain female shares across various career levels in natural science / academia. In the latter, a semi-structured interview guide loosely based on Cornelia Helfferich served as an interview blueprint for qualitative research focusing on personal experiences related to career choice and career advancement in natural science / academia. Taken together, the presented methods can be understood as valuable tools to obtain comprehensive information on female shares (**Hypothesis 1**) as well as individual perspectives in natural science / academia (**Hypothesis 2**). The respective findings of the data report analysis (**Hypothesis 1**) and international qualitative study (**Hypothesis 2**) are presented in the following **Section** 4.*

Results

4

Based on the above elaborated methods, **Chapter 4** presents the obtained results. The reader gains a detailed overview of single findings contributing to either **Hypothesis 1** or **Hypothesis 2**. In the former, female shares across various career levels in natural sciences / academia are unveiled. To ensure significance, respective statistics from Germany, the European Union and the United States of America are evaluated. In the latter, respective findings of the international qualitative study are revealed. Amongst others, 1) the personal and professional background of study participants, 2) their future aspirations and perception of natural science / academia as well as 3) perceived reason to stay or leave are depicted. In addition, individual believes on female underrepresentation are described. In a nutshell, the here presented experience-based results are the foundation to later rationally discuss **Hypothesis 2**.

Supplementary Information The online version contains supplementary material available at https://doi.org/10.1007/978-3-658-43086-3_4.

4.1   Hypothesis 1: Gender Distribution in Natural Science

4.1.1   Gender Distribution in Natural Science – Germany

On the grand scale, women held 27.2% of all professorships across all disciplines in Germany in 2021 (Statistisches Bundesamt (Destatis), 2022, p. 18–19). In the same year, women accounted for 21.5% of professor in natural science and mathematics (Statistisches Bundesamt (Destatis), 2022, p. 45 f.). To evaluate current gender distribution across career levels in natural science / academia, the report *Bundesbericht Wissenschaftlicher Nachwuchs 2021* was studied (Krabel et al., 2021). This national report summarises statistical data and research findings on doctoral students and graduates across all 16 federal states of Germany (Krabel et al., 2021, p. 106–107). Amongst others, it highlights female shares at various career levels in natural science / academia. **Figure 4.1a** illustrates the respective findings for 2018. Interestingly, nearly 50% of Master of Science graduates as well as PhD students were female. The share slightly decreases by 2 percentage points at PhD level. In short, women received of 45% of all doctoral titles in natural science in 2018. In the same year, 43% of newly appointed junior professors were women. However, women obtained less than one third of all habilitations. When looking at newly appointed W2 and W3 professorships (W3 being the better funded position), only 34% and 27% obtained a professorship in 2018, respectively.

When replotting gender distributions against increasing career levels as a *line with marks* chart, a scissor-shaped curve is obtained. This is illustrated in **Figures 4.1b**.

In addition, it is noteworthy that, excluding professors, 98% of all scientific full-time staff under the age 35 years were employed temporarily. Further, 92% of personnel under the age of 45 years had fixed-term contracts. On average, contract durations amounted to 22 months for PhD students and 28 months for postdocs (Krabel et al., 2021, p. 108).

a

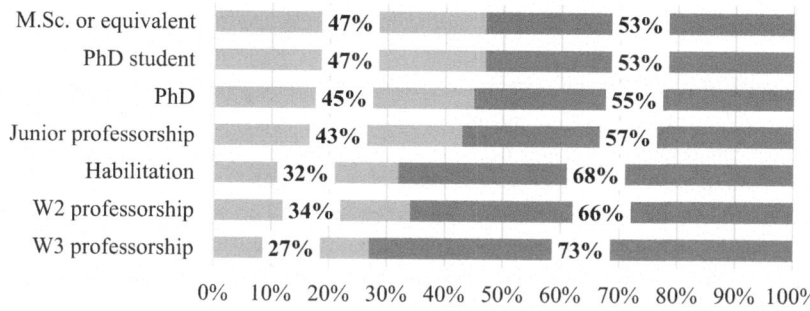

**Gender distribution across career
levels in natural science in 2018**

b

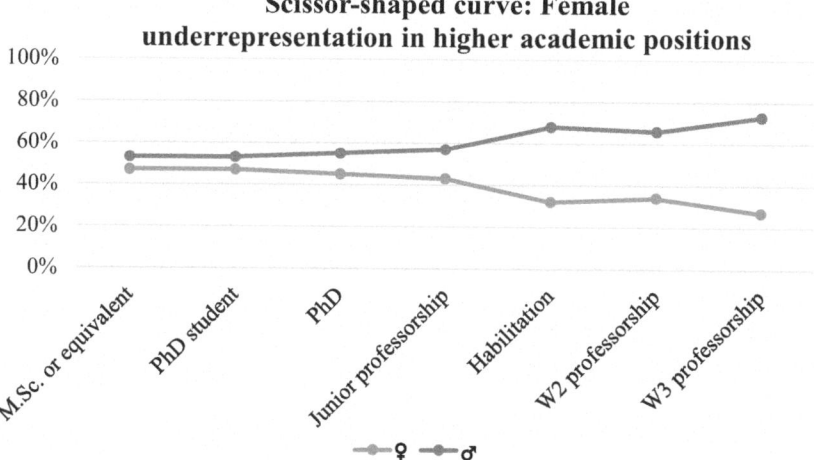

**Scissor-shaped curve: Female
underrepresentation in higher academic positions**

Figure 4.1   a) 100% stacked bar chart and **b)** line with markers chart of gender distribution across career levels in natural science in Germany (2018). (Source: Data for 1.) **M.Sc. or equivalent**: Statistisches Bundesamt (2020): Prüfungen an Hochschulen, Sonderauswertung, Wiesbaden; 2.) **PhD students**: *Statistisches Bundesamt (2019): Promovierendenstatistik: Analyse zu Vollständigkeit und Qualität der zweiten Erhebung 2018, Wiesbaden*; 3.) **PhD**: *Statistisches Bundesamt (2019): Prüfungen an Hochschulen—Fachserie 11, Reihe 4.2, Wiesbaden*; 4.) **Habilitation**: *Statistisches Bundesamt (2019): Personal an Hochschulen— Fachserie 11, Reihe 4.4, Wiesbaden*; 5.) **Junior and W2 / W3 professorship**: *Statistisches Bundesamt (2020): Personal an Hochschulen, Sonderauswertung, Wiesbaden*. Reference: (Krabel et al., 2021, S. 108). Author's own graph.)

4.1.2   Gender Distribution in Natural Science – Europe

To analyse gender distributions on a European scale, respective data sets from EU countries were evaluated. The report *She Figures* provides cross-European, comparable statistics on female shares in research and innovation (European Commission & Innovation Directorate-General for Research, 2021). Based on these statistics, at EU level, women account for 54% and 59% of B.Sc. and M.Sc. (ISCED 6 and 7) students and graduates in 2018, respectively. Please note that the data referred to *typical academic careers*. However, the exact definition of *typical academic careers* was not provided. Consequently, the data is not directly comparable to the German statistic presented above. The EU average of female shares of PhD (ISCED 8) students and graduates was 52%. In addition, women accounted for 47% grade C staff (first position after PhD), 40% of more senior grade B staff but only 26% of grade A staff (single highest position, e.g., full professorship position) (European Commission & Innovation Directorate-General for Research, 2021, p. 21 ff.). **Figures 4.2a** illustrates the elaborated gender distributions. In short, at the EU level, women represented nearly half of the academic staff on average in 2018. However, only one-in-four grade A positions was held by a woman. Among heads of institutions in the higher education sector, women accounted for 23% in 2019 (European Commission & Innovation Directorate-General for Research, 2021, p. 175 ff.). In **Figures 4.2b** the presented data was replotted as a *line with marks* chart. In addition, values for respective gender shares from 2015 were included. Remarkably, female shares across ISCED 6, 7 and 8 students and graduate were identical in 2015 and 2018. However, women made slight gains in grade C (2 percentage points), B (1 percentage point) and A (1 percentage point). In a nutshell, a scissor-shaped curve is obtained when plotting gender distribution against increasing career levels.

a

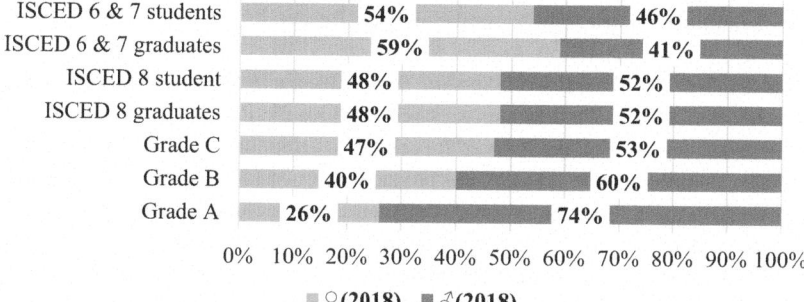

b

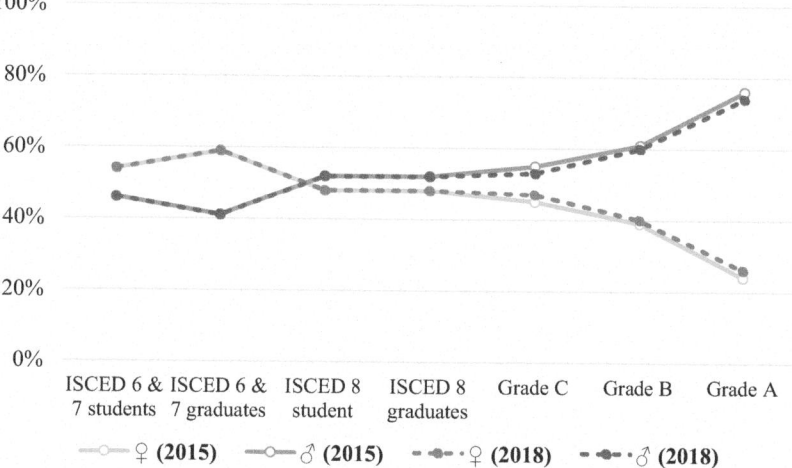

Figure 4.2 **a)** 100% stacked bar chart and **b)** line with markers chart of gender distribution across career levels in natural science in the EU (2018). (Source: Women in Science database, DG Research and Innovation—T1_questionnaires, Education Statistics (online data codes: educ_uoe_enrt03, educ_uoe_grad02) Reference: (European Commission & Innovation Directorate-General for Research, 2021, p. 176 ff.). Author's own graph)

4.1.3   Gender Distribution in Natural Science – U.S.

To further evaluate female shares in natural science / academia, statistics from the Unites States of America—a leading country in natural-sciences research ("The Ten Leading Countries in Natural-Sciences Research," 2020)—were analysed. Amongst others, the report *Annual Report on the Workforce, US Equal Employment Opportunity Commission, Special Topic: Women in STEM* (Office of Federal Operations, 2019) as well as statistics provided by the Pew Research Centre (Fry et al., 2021) were evaluated. Overall, gender distributions were listed for STEM degrees, degrees in life science and physical science (see **Figure 4.3**). It should be noted that STEM is the umbrella term and thus includes, amongst others, the respective data for natural science (i.e., life and physical science.).

Strikingly, women obtained the majority of M.Sc. in STEM and life science in the school year 2017–18. Only in physical science, the female shares in M.Sc. graduates were below 40%. However, whilst female shares significantly declined from M.Sc. to PhD in STEM and life science, gender distributions in physical science stayed nearly constant (Fry et al., 2021).

Further, women accounted for 50% of the STEM workforce (including both academia and industry). In detail, women made up 48% and 40% of staff in life and physical science. In particular, female shares in physical science degrees and workforce aligned well. **Figure 4.4** summarises these findings. In contrast, women only accounted for 27% of professors in STEM in 2021 (see **Figure 4.5**) and thus, lying far below the average female shares in higher education and STEM jobs.

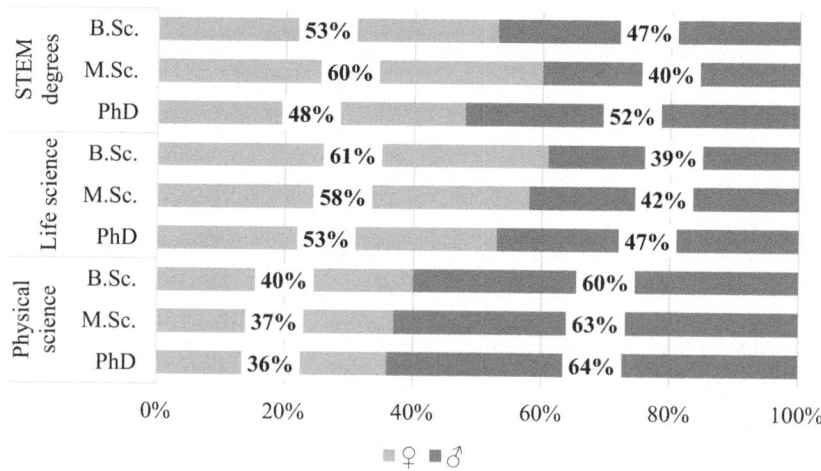

Figure 4.3 Degrees awarded in STEM based on U.S. citizens and permanent residents in 2017–18. (Source: U.S. Department of Education, National Center for Education Statistics, Integrated Postsecondary Education Data System analysed using the National Center for Science and Engineering Statistics Interactive Data Tool, 2017–18 school year. Reference: (Fry et al., 2021). Author's own graph)

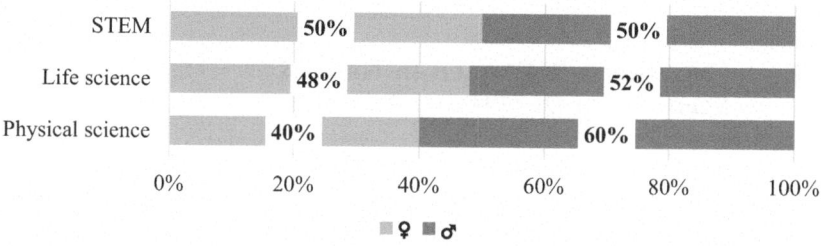

Figure 4.4 Overall share of women in STEM jobs in the U.S. (Source: Pew Research Center analysis from 2017–19 American Community Survey (IPUMS). Reference: (Fry et al., 2021). Author's own graph)

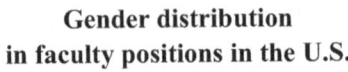

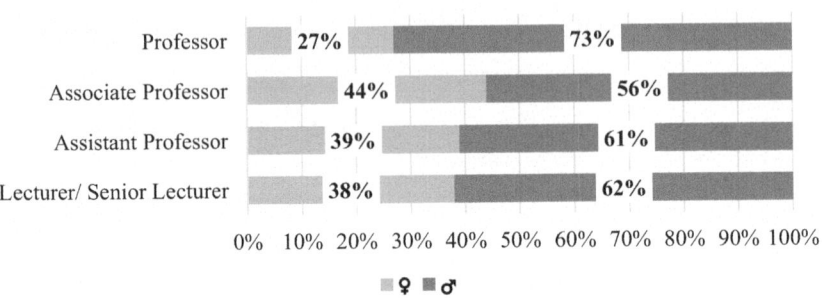

Figure 4.5 Gender distribution in faculty positions in the U.S. (Source: Office of Institutional Research (2021). Reference: (Mehta, 2022). Author's own graph)

4.1.4   Key Takeaway: Gender Distribution in Natural Science

In a nutshell, gender parity has nearly been reached across higher education levels in Germany, the EU and the U.S. However, female share declined with increasing career level in natural science / academia. It is particularly striking that women account for less than 30% of professors in Germany, the EU and U.S. Thus, less than one-in-three professors is a woman. As a result, a scissor-shaped curve can be obtained when plotting gender distributions across career levels in natural science / academia. The here-presented findings will be interpreted and discussed in **Section 5.1**.

4.2   Hypothesis 2: Women In Natural Science – a Qualitative Analysis

Between the 20th of May and 15th of September, 26 female scientists participated in the here presented qualitative study. Every woman was interviewed in a one-on-one interview via video call. On average, the actual interview (without introduction, technical check, etc.) took 42 min with the shortest amounting to 23 min and the longest to 74 min. In total, 18.4 h of recording were generated. In this section the respective findings will be presented. To this end, **Section 4.2** is subdivided into four subsections with thematical emphasis on:

A) Characterisation of study participants,
B) Future aspirations and perception of natural science / academia,
C) Reasons to stay in / leave natural science / academia,
D) *Goal congruity theory* and female underrepresentation in science / academia.

4.2.1 Characterisation of Study Participants

All study participants were women with a background in natural science and engineering. In detail, 10 participants had a background in biology, 6 in chemistry, 5 in biochemistry, 3 in physics, 1 in engineering and 1 in veterinary medicine (see **Figure 4.6**).

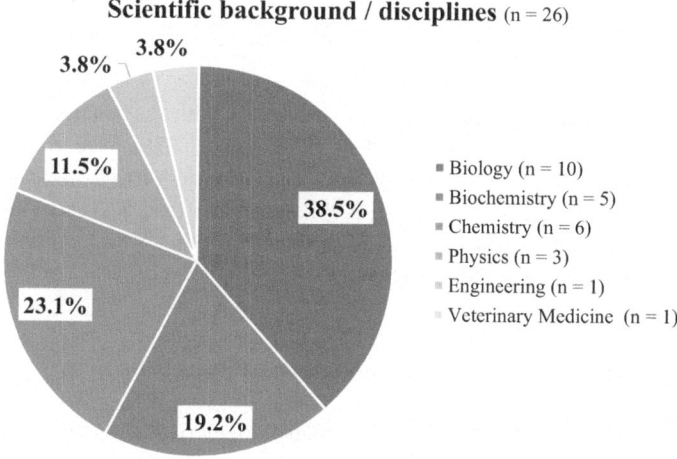

Figure 4.6 Scientific background / disciplines of study participants

All participants held at least a PhD title in their area of science. The female scientists were specialised in vastly differing research fields; in total, they represented 17 research fields. **Figure 4.7** depicts all fields and the respective shares of women. Approx. 19.2% focused on cancer biology (5 / 26 participants), a 11.5% on organic synthesis (3 / 26 participants) and approx. 7.8% on astronomy (2 / 26 participants). The other 14 research fields were represented by one woman each.

Out of all, 73.1% of researchers held temporary (short-term, e.g., 3 moths to 5 years) contracts. The other 7 had a contract of employment of indefinite duration (see **Figure 4.8a**); out of them, 2 were teaching staff, 1 head of a core facility (comparable to a group leader position), 1 research assistant, 1 assistant professor and two held positions outside of academia (1 in industry and 1 in scientific consulting, see **Figure 4.8b**). **Figure 4.8c** illustrates the types of temporary position held by temporarily employed participants. The majority (68.4%, 13 / 19 participants) were employed as postdocs, 2 as junior group leader and one each as habilitation candidate, independent researcher, astronomer. Another woman was employed at a governmental research institute (which is not a classical academic position).

The average age of participants was 35.8 years with the youngest being 27 and the oldest 52 years (see **Figure 4.9**). In summary, 7.7% were younger than 30 years (2 / 26), 73.1% were between 30 and 39 years (19 / 26), 15.4% were between 40 and 49 years (4 / 26) and 3.9% were older than 50 years (1 / 26).

Study participants originated from 13 different nations across 5 continents (see **Figure 4.10a**). Their distribution across respective continents is listed below:

Asia:	2 (India),
North America:	6 (Canada, Trinidad and Tobago, USA),
South America:	1 (Brazil),
Europe:	15 (France, Germany, Netherlands, Poland, Portugal, Spain, UK),
Australia / Oceania:	2 (New Zealand).

Research fields of study participants (n = 26)

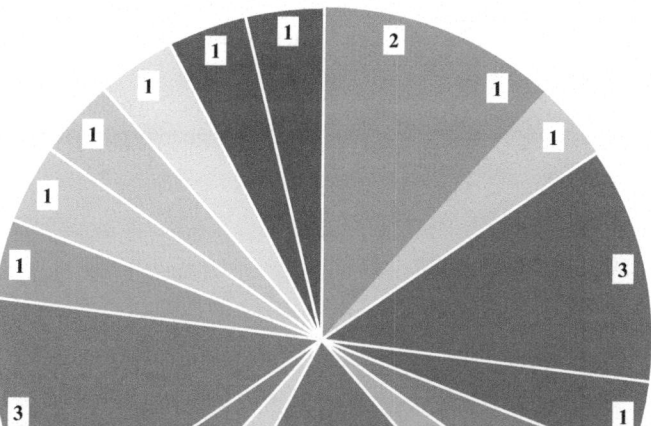

- ▪ Astronomy
- ▪ Inorganic chemistry
- ▪ Biomedical Microtechnology
- ▪ Medicinal Chemistry / Oncology
- ▪ Neurobiology
- ▪ Tropical diseases
- ▪ Computational Biochemistry
- ▪ Plant biology
- ▪ Microfluidics

- ▪ Molecular biology / spectroscopy
- ▪ Organic Chemistry
- ▪ Drug development
- ▪ Cancer Biology
- ▪ Ageing Biology
- ▪ Metabolomics
- ▪ Microbiology
- ▪ Corral biology

Figure 4.7   Research fields of study participants

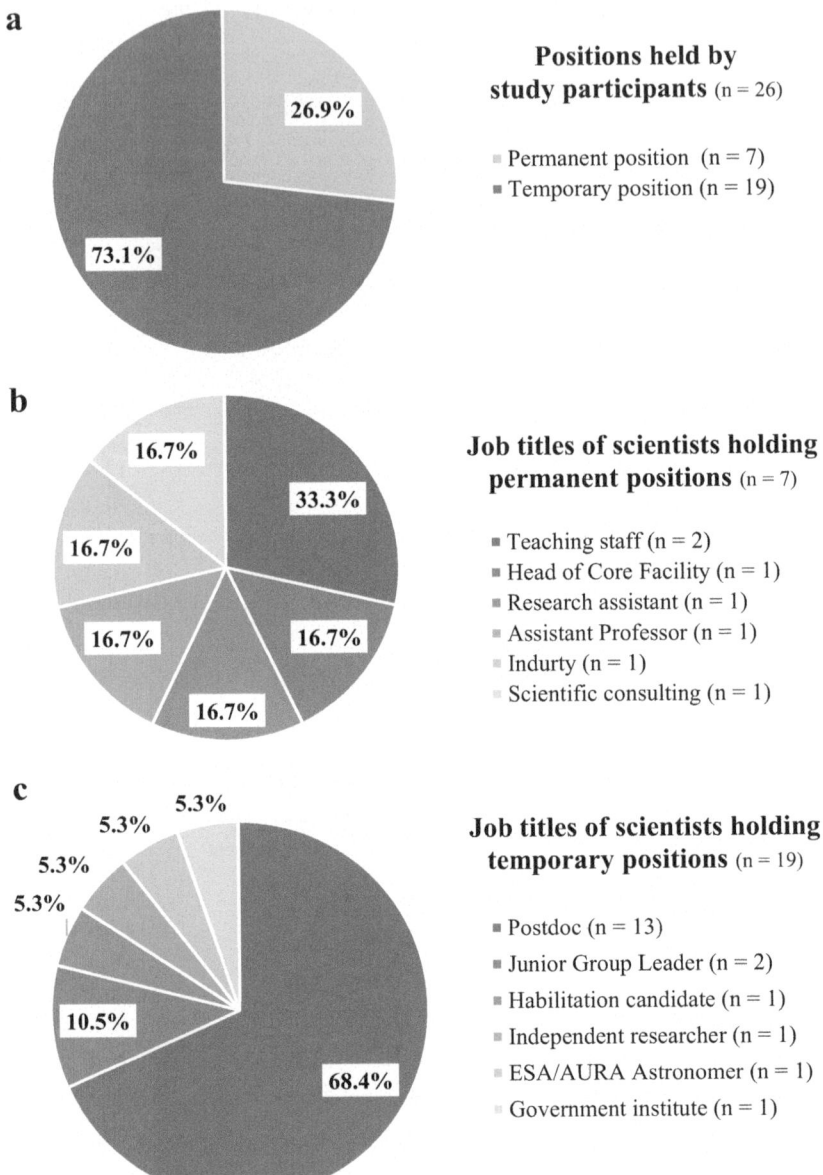

a

**Positions held by
study participants** (n = 26)

- Permanent position (n = 7)
- Temporary position (n = 19)

b

**Job titles of scientists holding
permanent positions** (n = 7)

- Teaching staff (n = 2)
- Head of Core Facility (n = 1)
- Research assistant (n = 1)
- Assistant Professor (n = 1)
- Indurty (n = 1)
- Scientific consulting (n = 1)

c

**Job titles of scientists holding
temporary positions** (n = 19)

- Postdoc (n = 13)
- Junior Group Leader (n = 2)
- Habilitation candidate (n = 1)
- Independent researcher (n = 1)
- ESA/AURA Astronomer (n = 1)
- Government institute (n = 1)

Figure 4.8   **a)** Employment type, **b)** job titles of indefinite and **c)** fixed-term employment

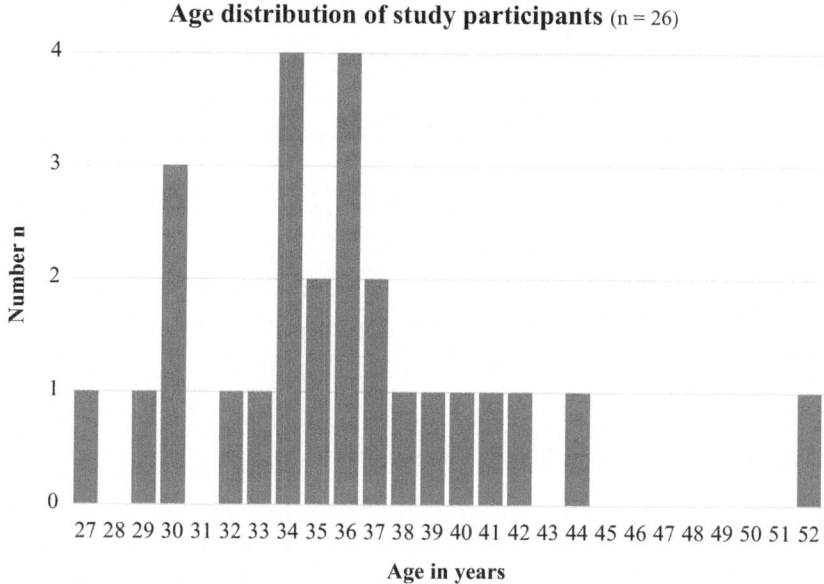

Figure 4.9 Age distribution of study participants

At the time of interview, the scientists were employed in 8 different nations across 4 continents:

North America: 6 (Canada, USA),
South America: 1 (Brazil),
Europe: 15 (Germany, Spain, UK),
Australia / Oceania: 4 (Australia, New Zealand).

Figure 4.10b depicts the individual countries of work affiliation.
Along their scientific career, the interviewed women have worked in 6 out of 7 continents (excluding Antarctica). Counting single U.S. states and countries of the UK separately, the scientists worked in 79 states and countries across the globe. **Figure 4.11a** depicts all academic workplaces and a comprehensive list including all relevant U.S. states and UK nations can be found in electronic supplementary material (EMS) on page 10. In summary, 8.9% of workplaces were in Asia, 2.5%

in Africa, 29.1% in North America, 1.3% in South America, 53.2% in Europe and 5.1% in Australia / Oceania.

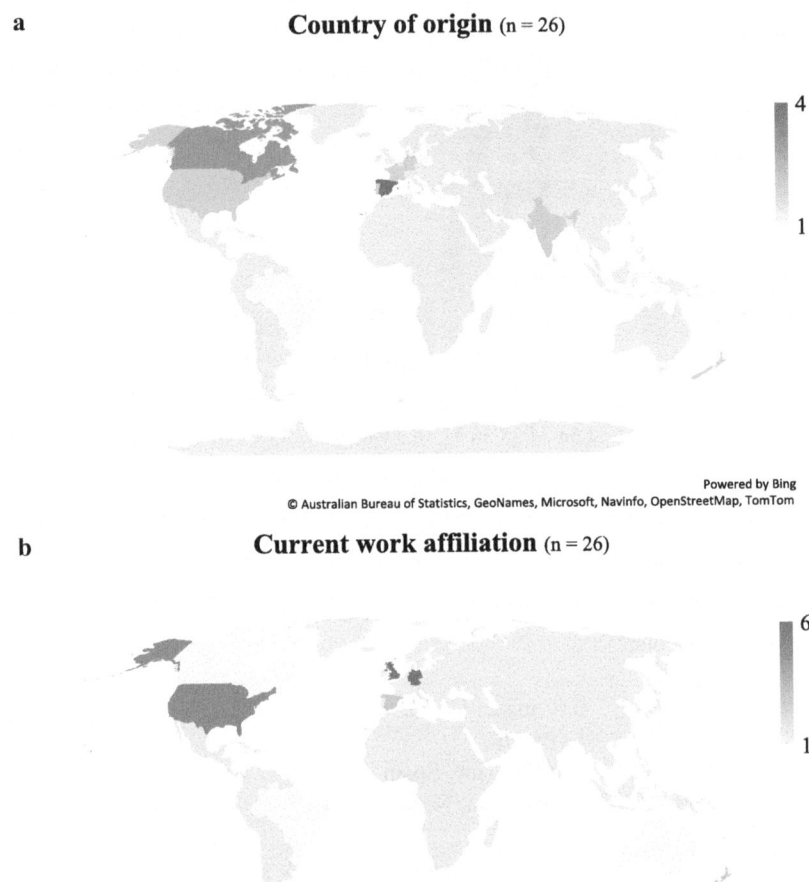

Figure 4.10   a) Country of origin and b) current work affiliation of study participants

More detailed, the scientists held positions in the following countries:

Asia: 7 (China, Japan, Laos, India, Indonesia, Saudi Arabia),
Africa: 2 (Kenya, Republic of The Gambia),
North America: 23 (Canada, USA),
South America: 1 (Brazil),
Europe: 42 (France, Germany, Netherlands, Poland, Portugal, Spain, UK),
Australia / Oceania: 4 (Australia, New Zealand).

Interestingly, this classification can be modified to represent political associations rather than continents. Accordingly, **Figure 4.11b** clusters the work experiences; here, 8.9% of academic positions were placed in Asia, 1.3% in South America, 39.2% in the European Union, 29.1% in the Commonwealth of Nations and 21.5% in the United States of America.

Explicitly, the researchers worked in the following countries:

Asia: 7 (China, Japan, Laos, India, Indonesia, Saudi Arabia),
South America: 1 (Brazil),
United States: 17 (USA—see **EMS**, p. 10),
European Union: 31 (France, Germany, Netherlands, Poland, Portugal, Spain),
Commonwealth: 23 (Australia, Canada, Kenya, New Zealand, Republic of The Gambia, UK – see **EMS**, p. 10).

In conclusion, study participants worked in 3 different countries / states on average (excluding potential inner-country / -state work relocations). In fact, 92.3% of participant relocated countries at least once, 57.6% at least twice and 42.3% at least three times during their academic career in natural science. Out of 26, only 2 participants had not relocated countries. The geographical and occupational relocation of workers is also referred to as labour mobility. The respective labour mobility of study participants is illustrated in **Figure 4.12**.

Regarding the personal background, 11 women stated to be married, 7 to be in a long-term partnership and 8 to be single (see **Figure 4.13a**). The vast majority was childless. 4 stated to have a child, 1 to have children. 2 of them had already left academia (including the one with children, see **Figure 4.13b**). Out of 21, 10 stated a desire to have children. 1 was uncertain. Another 10 did not desire children; 40% of them were above the age of 36 (see **Figure 4.13c**).

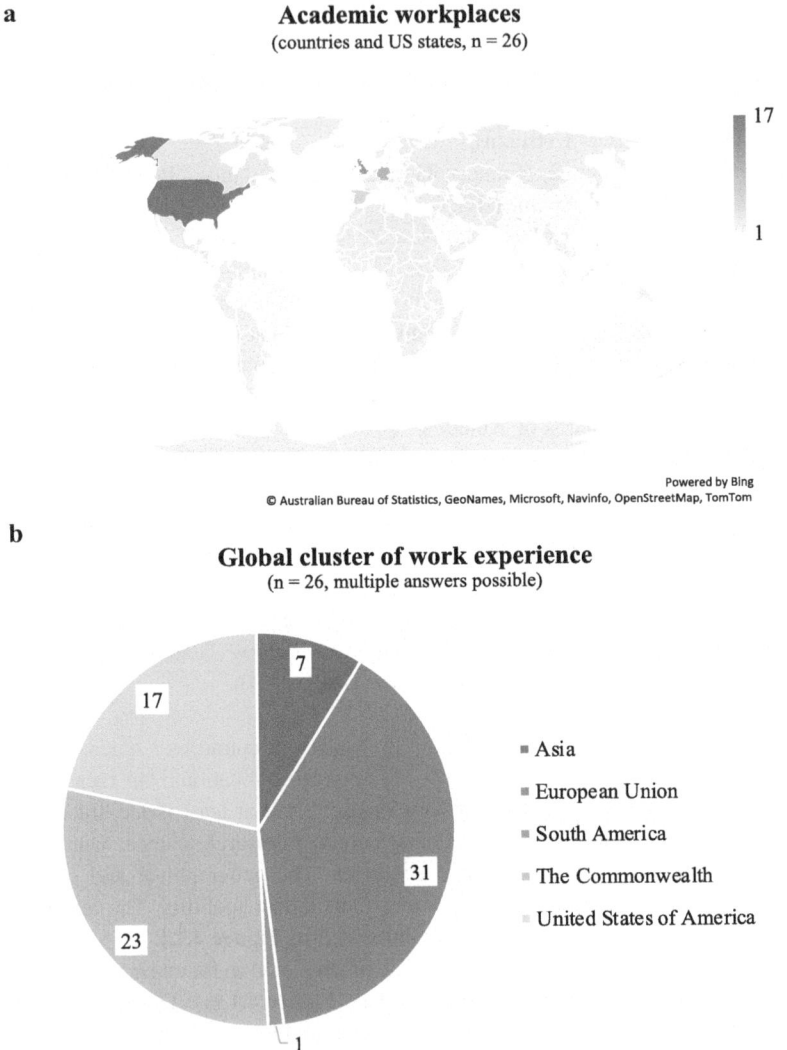

a

Academic workplaces
(countries and US states, n = 26)

17

1

Powered by Bing
© Australian Bureau of Statistics, GeoNames, Microsoft, Navinfo, OpenStreetMap, TomTom

b

Global cluster of work experience
(n = 26, multiple answers possible)

- Asia
- European Union
- South America
- The Commonwealth
- United States of America

Figure 4.11 **a)** Academic workplaces and **b)** global cluster of work experience of study participants

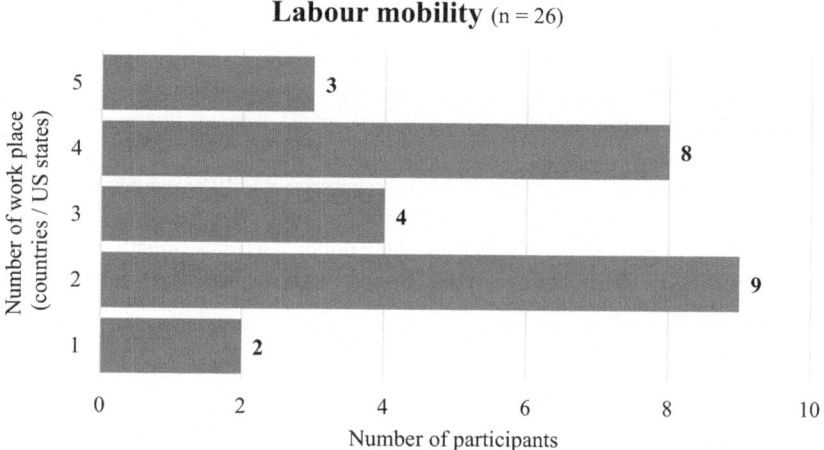

Figure 4.12 Labour mobility of study participants

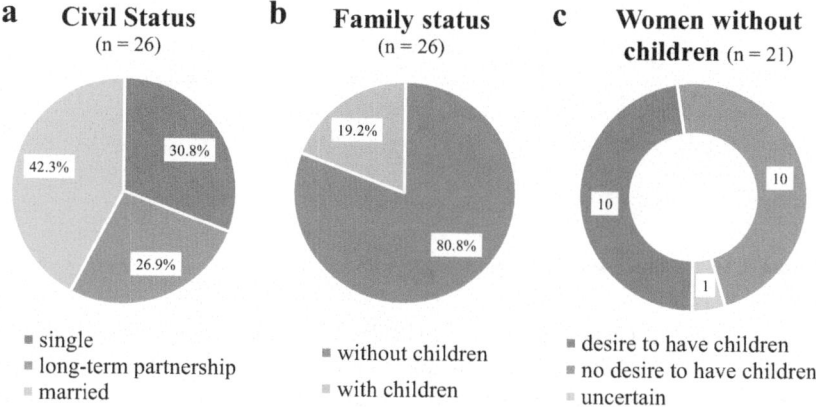

Figure 4.13   **a)** Civil status, **b)** family status and **c)** women without children and their respective desire to have children

4.2.2   Future Aspirations and Perception of Natural Science / Academia

When participants were asked to state their future aspirations, the vast majority wished to continue in academia. In fact, 19 of 23 women, that were directly employed in natural science / academia, wanted to progress in the academic career path. From the remaining 4, 2 researchers wanted to leave academia and 2 were uncertain. Another 2 women had indefinitely left academia. 1 scientist currently employed by a government institute, thus, not directly employed in academia, stated to be uncertain. **Figure 4.14** depicts the representative percentages based on all 26 participants.

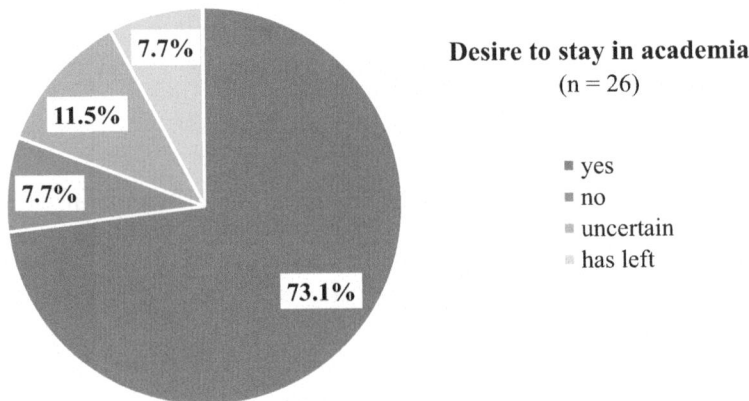

Figure 4.14   Desire to stay in natural science / academia

In response to leading question 2 (*As a Postdoc/ junior group leader you hold a great track record in academia. Would you please share your future aspirations with me?*, see **EMS**, p. 6), the most common answer was *to obtain a professorship* and / or a *tenure track* and / or a *PI position*. In total, these job titles were mentioned 17 times and accounted to statements from 14 women. (Please note that multiple answers were possible.) Interestingly 10 out of these 14 (≙ 71.4%) solely named a / several position(s) as their future aspiration(s). The other 4 also named research-associated goals (i.e., *moving to a different field of science, being a good scientist and mentor, mission-driven work*). The second most common answer was *uncertain* (4 women), followed by *implementation-based work* and *doing things of value* (each 3 women). In addition, 4 women wanted to either shift to or advance in industry. A comprehensive overview of all responses is depicted in **Figure 4.15a**.

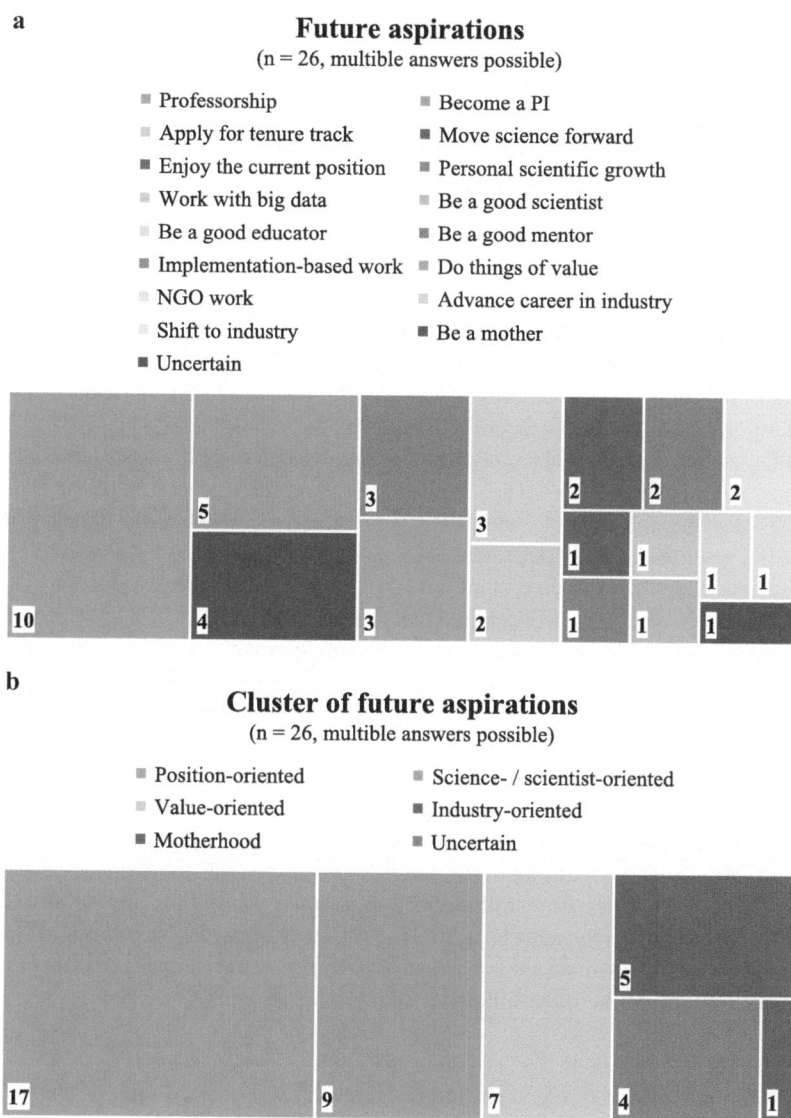

Figure 4.15   a) List and b) cluster of future aspirations of study participants

To improve data significance, individual answers of participants were clustered into 6 categories: 1) *position-oriented*, 2) *science- / scientist-oriented*, 3) *value-oriented*, 4) *industry-oriented*, 5) *motherhood* and 6) *uncertain*. The respective allocation of response can be seen in **EMS**. **Figures 4.16b** summarises the result; academic positions were mentioned 17 times, science-oriented goals 9 times, value-oriented goals 7 times and industry-oriented goals 5 times as future aspiration(s).

From the 23 women actively working natural science / academia, 14 aspired a higher position,

4 aspired a job in industry and / or NGO (currently working as postdoc / teaching staff), 2 wanted to continue in their current position (i.e., as astronomer, head core facility) and 3 were uncertain (currently working as postdoc / independent researcher). In total, only 1 woman who held a permanent position was looking to transition from academia to industry.

Please note that the data sets *Desire to stay* (**Figure 4.14**) in academia and *Future aspirations* (**Figure 4.15a**) are based on two independent questions. Thus, results deviate; for example, at least 3 women expressed the explicit desire to stay but were uncertain about future aspirations in natural science / academia. In a nutshell, 82.6% of women currently working in natural science / academia (19 / 23) wanted to continue the academic career path. Out of them, 73.7% (14 / 19) desired a group leader position / tenure track / professorship, 10.5% (2 / 19) were content with their position and 15.8% (3 / 19) uncertain about future aspirations.

Subsequently, study participants were asked to describe the scientific community with 3—5 attributes (leading question 3, see **EMS**, p. 6). This interview question intended to unveil individual perceptions of the science community. In total, 71 different attributes were named; among them, *collaborative* (10 / 26 participants) and *competitive* (8 / 26) were named most frequently. *People-depen*dent, *open-minded* (each 5 / 26) and *community* (4 / 26) were also mentioned several times. These attributes were followed by *innovative, passionate* and *financially unsustainable / underfunded* (each 3 / 26). All other attributes were only named once or twice. To increase the explanatory power of selected attributes, all were allocated to one of the three following codes and subsequently scored:

A) Positive (Score: + 1; e.g., *collaborative, supportive, motivated*),
B) Neutral (Score: ± 0; e.g., *competitive, people-dependent, driven*),
C) Negative (Score: −1; e.g., *abusive, career blocking, unethical*).

A comprehensive list and the respective scores can be found in **EMS** on page 13. In general, all attributes promoting a positive work environment were scored as

+ 1. All that contributed to a toxic work environment (i.e., *aggressive, obsessive, precarious*) were scored as − 1. All others that were not directly associated with a positive or negative impact were classified as neutral with a score of ± 0 (e.g., *driven, ambitious, challenging*).

After scoring single attributes for each participant (see **EMS**, p. 15), a total score was determined by adding up all respective values. The results are illustrated in **Figure 4.16**.

Interview code	#01	#02	#03	#04	#05	#06	#07
Individual score of listed attributes	1	0	0	-1	1	0	1
	1	-1	0	1	0	1	1
	0	-1	-1		1		1
		0			0		-1
		-1			1		
Total score	2	-3	1	0	3	1	2

Interview code	#08	#09	#10	#11	#12	#13	#14
Individual score of listed attributes	0	1	0	1	0	1	0
	-1	1	1	0	1	1	1
	1	1	0	0	1	1	0
	1			-1	1		0
	1				0		
					-1		
Total score	3	3	1	0	2	3	1

Interview code	#15	#16	#17	#18	#19	#20	#21
Individual score of listed attributes	-1	1	1	1	-1	1	1
	-1	0	-1	1	0	0	1
	-1	1		-1	-1	1	1
	-1	1		0	1	1	1
	1			1	1		0
	1				1		-1
					1		
					1		
					1		
Total score	-2	3	0	2	4	3	3

Interview code	#22	#23	#24	#25	#26
Individual score of listed attributes	1	1	1	1	0
	1	-1	1	1	0
		-1	-1	1	0
		-1	-1	0	-1
		1	-1		-1
Total score	2	-1	-1	3	-2

Colour coding system		
Individual score	Total score	Interpretation
1	4	Positive perception
	3	
	2	
	1	
0	0	Neutral perception
-1	-1	Negative perception
	-2	
	-3	

Figure 4.16 Individual perception of the scientific community

In fact, **Figure 4.16** can be seen as a heat map of individual perceptions of the study participants. Each black-framed square represents a scientist and her respective scores. The sum of these is shown in the bold-framed total score row. A colour coding system visualises individual and total scores. Remarkably, 17 participants reached a score above zero. Thus, in total, positive attributes predominated the individual perceptions. In contrast, negative impressions overweighed in 6 participants. For 3 women, positive and negative attributes cancelled each other out to a score of ± 0 (see **Figure 4.17a**). **Figure 4.17b** illustrates the deduced perception based on the score values. In a nutshell, more than a two thirds majority described a positive perception of the experienced scientific community. It should be noted that study participants were able to define whom they counted as part of scientific community. Thus, every woman described their own perception for their experienced scientific community. The focus on an experience-based description was deliberate. The respective interpretation will be discussed in **Section 5.2.2**.

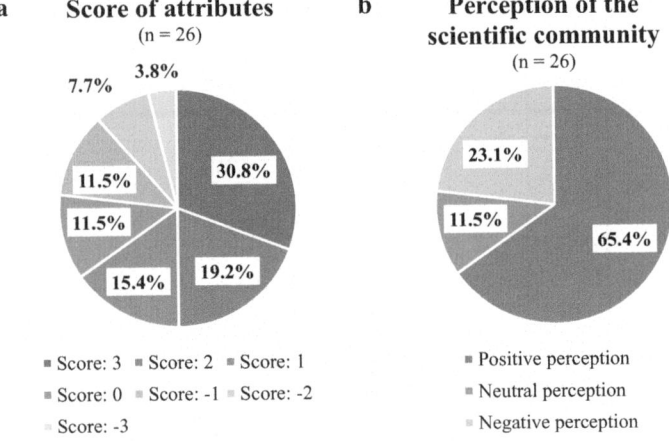

Figure 4.17   **a)** Score of attributes and **b)** score-based perception of the scientific community

4.2.3   Reasons to Stay in / Leave Natural Science / Academia

The next interview question was designed to identify perceived reasons to either stay in or leave natural science / academia:

> *Many scientists have to face the choice to either stay in academia / or leave academia. From a personal perspective, could you name relevant factors that may affect this choice?*

Women listed various reasons and quickly elaborated them. In summary, only 23 women named reasons to stay in natural science / academia. In total, 8 aspects were named across all participants. The following **Figure 4.16** depicts all mentioned reasons and their respective distribution. Please note that participants were able give several answers.

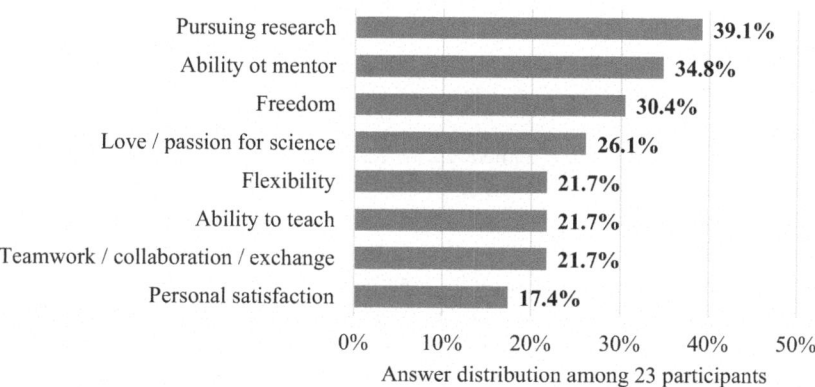

Reasons to stay in natural science / academia
(n = 23, multiple answers possible)

Answer distribution among 23 participants

Figure 4.18   Reasons to stay in natural science / academia

Most commonly, women named *pursuing research* (9 / 23), *the ability to mentor* (8 / 23) and *the personal and scientific freedom* (7 / 23) as reasons why one would stay in natural science / academia. In general, it is noteworthy that a total of 18 women listed community-based activities (i.e., *mentoring* and / or *teaching* and / or *teamwork / collaboration / exchange*) as positive aspects in natural science / academia.

As a counterpart to **Figure 4.18**, **Figure 4.19a** illustrates listed reasons that promote people to opt out of natural science / academia. In contrast to a readily comprehensible number of reasons to stay, a plethora of reasons to leave were listed by the participants. Individual aspects were diverse. On average women name 4 different reasons. To create for meaningful overview, all reasons listed were reviewed. Subsequently, similar ones were clustered and coded using one of the following the codes.

A) Family commitment,
B) Frustration,
C) Lack of funding,
D) Lack of positions,
E) Lack of stability,
F) Life choices,
G) Mental health,
H) Pressure,
 I) Required mobility,
J) Salary,
K) Time commitment,
L) Work environment,
M) Work-life-balance.

In total, study participants named 98 aspects that were coded and graphically illustrated accordingly (see **Figure 4.19a**). In first place, *time commitment* and *pressure* (each 12 / 26), in second, [low] *salary* and *lack of stability* (each 11 / 26) and in third, *lack of funding* were named. In particular, the categories *pressure* (total count: 12) and *lack of stability* (total count: 11) can further be differentiated into subcodes. For the former, the respective stated *pressure to succeed* (5 / 12) and / or *pressure to publish* (4 / 12) and / or *financial pressure* (3 / 12) and / or *societal pressure to fulfil gender roles* (2 / 12) as major constraints. For the latter, 8 women specified *lack of job stability* and 1 named *lack of financial stability* as main contributors to the general *lack of stability* in natural science / academia. Interestingly, *family commitment* was only mentioned 6 times. In contrast, monetary aspects (i.e., *salary, lack of funding, financial pressure* and *lack of financial stability*) were named 25 times thus accounting for 25.5% (25 / 98) of all listed aspects. Further, to identify superordinate themes, the 13 codes listed above were reduced to 5 categories (see **Figure 4.19b**): 1) *uncertainty*, 2) *mental wellbeing*, 3) *workload*, 4) *salary* and 5) *individual choices*. The combination of financial instability (i.e., *funding*), job instability, lack of positions and required mobility can be summarised as *uncertainty*. Amongst all aspects listed, one third accounted for *uncertainty* (33 / 98). Runner-ups were *mental wellbeing* (i.e., *pressure, mental health,* work *environment* and *frustration*) and *workload* (i.e., *time commitment, work-life-balance, family commitment*), each were named 25 times. In total, 84.7% (83 / 98) of perceived reasons to leave natural science / academia belong to either of these 3 categories.

a

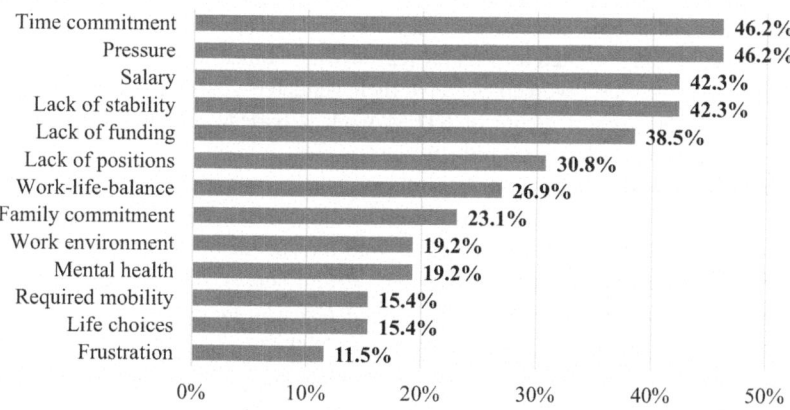

b

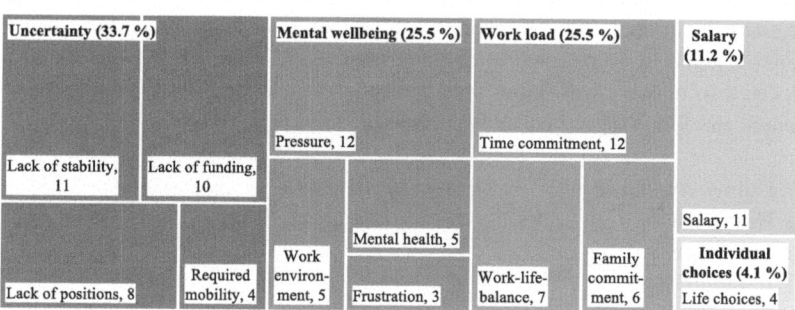

Figure 4.19 a) List and b) cluster of reasons to leave natural science / academia

4.2.4   *Goal Congruity Theory* and Female Underrepresentation in Science / Academia

After having discussed perceived reasons to leave natural science / academia, all study participants were confronted with the following statement (orally and in written form).

Women choose to leave natural science to fulfil goals that they perceive to be mismatched with the nature of academia.

This statement is based on the *goal congruity theory* introduced by Diekman *et al.* (Diekman et al., 2010, 2020)—a framework already introduced in the introduction and commonly quoted as a major reason why women leave natural science / academia. Simply put, women perceive their gendered goals/values to be incongruent with the general nature of STEM work (Charlesworth & Banaji, 2019). Thus, they decide to leave natural science.

In this qualitative study, study participants were asked to comment on the statement above without being introduced to the theoretical basis (i.e., *goal congruity theory*). Remarkably, 50% agreed with the statement. 3 of them explicitly stated that they agreed but for all sexes (i.e., ***People*** *choose to leave natural science to fulfil goals that they perceive to be mismatched with the nature of academia.*). Another 2 women partially agreed with the statement. 5 participants disagreed partially or mostly. 6 women disagreed. In a nutshell, 57.5% support the presented statement. **Figure 4.20** summaries the respective findings.

Next, women were asked to share their individual beliefs on female underrepresentation in natural science / academia. Thus, this question aimed to identify the individual beliefs (i.e., perception) rather than hard facts. Individual answers were clustered in accordance to common themes. In total, 65 answers could be allocated to clusters with more than 1 response each. Based on the represented themes, the following 10 codes were chosen.

A) Family commitment
B) Hiring bias
C) Imbalance
D) Instability
E) Lack of role models
F) Lack of self-confidence
G) Lack of support
H) Lack of visibility
 I) Stereotypes / gender roles
J) Toxic environment

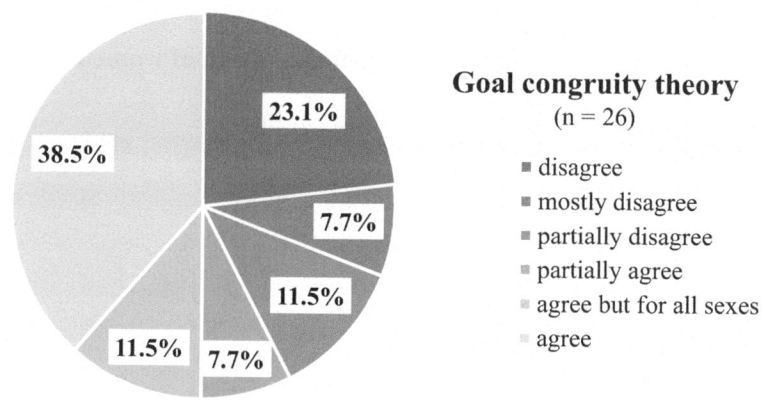

Figure 4.20 *Goal congruity theory*: participants' comments

Figure 4.21a illustrates their respective distributions. Nearly two thirds believed that stereotypes / gender roles are causes of female underrepresentation in natural science / academia. *Hiring bias* and [power] *imbalance* (i.e., overrepresentation of men in decision-making positions) were named second and third most commonly. Participants also stated *lack of support* (e.g., childcare, support from supervisors), *toxic work environment, lack of self-confidence, lack of role models* and *family commitment* as reasons why women are underrepresented in natural science / academia. It seems noteworthy that 5 women stated that women are less confident about their personal skills than men and thus miss out because of undervaluation. This phenomenon is literature-known and referred to as *gender confidence gap* (Sterling et al., 2020, see **Section** 1.4.3).

To increase explanatory power, the 10 codes were clustered in to 4 categories: 1) *antiquated views*, 2) *underrepresentation*, 3) *working conditions* and 4) *individual aspects*. Their respective distribution can be seen in **Figure 4.21b**. In short, 40% of answers correlated to *antiquated views* (e.g., historical role models and the idea that women are less hireable because they will have children). 24.6% of responses corresponded to [female] *underrepresentation* in positions with power. As elaborated above, female underrepresentation in higher positions is known as vertical segregation.

In detail, participants postulated that men tend to hire / promote men (i.e., implicit or explicit hiring bias). Thus, male overrepresentation in positions with hiring power facilitates large male shares. Male dominance in higher positions also contributes to greater visibility. Accordingly, visibility of women is reduced.

a

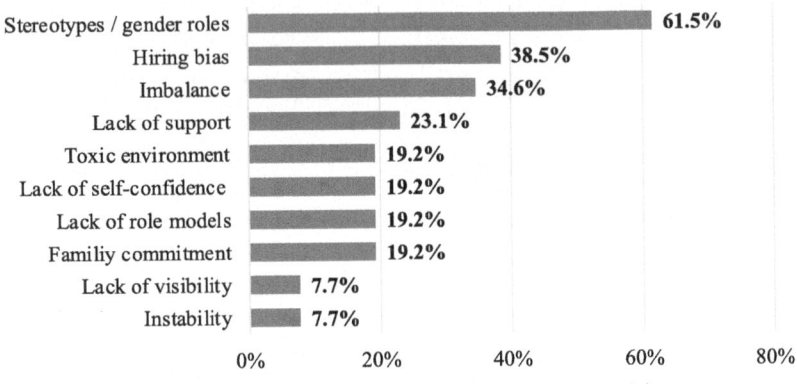

Individual beliefs: Underrepresentation of women
(n = 26, multiple answers possible)

b

Clusteres individual beliefs: Underrepresentation of women
(n = 26, multiple answers possible)

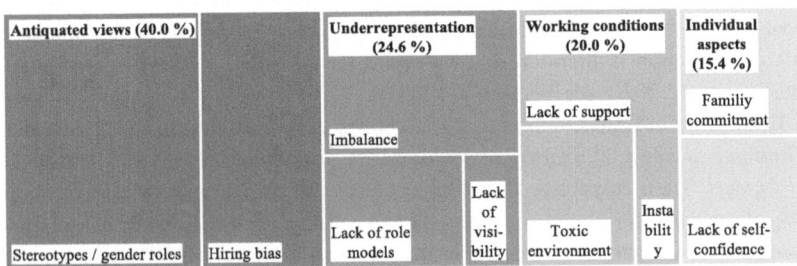

Figure 4.21 a) List and b) cluster of individual beliefs on the underrepresentation of women in natural science / academia

Another 20% of answers stressed *working conditions* as a major constraint. Finally, 15.4% indicated towards *individual aspects* such as *personal undervaluation* and *family commitment*. In summary, study participants believed that *antiquated views* and female *underrepresentation* majorly contribute to gender segregation in natural science / academia.

*In **Chapter 4**, extensive data to address **Hypothesis 1** and **Hypothesis 2** were presented. For the former (**Hypothesis 1**), statistics from 2015-21 unveiled near-gender parity in early career stages in Germany, the EU and USA. However, female shares drastically declined with increasing career levels in natural science. In particular, professorships were predominantly held by men across all evaluated regions. Thus, the presented results revealed male overrepresentation in decision-making positions. For the latter (**Hypothesis 2**), diverse backgrounds of study participants were revealed; in total, participants worked in 17 different fields, had current work affiliations across 4 continents and had worked in 79 countries and U.S. states across 6 continents. The vast majority of women stated a desire to stay and was working towards a permanent position in natural science / academia. Amongst others, the scientific community was perceived as collaborative, competitive and open-minded. Pursuing research, the ability to mentor and freedom were the top reasons to stay. Time commitment, pressure and salaries were the most common reasons to leave. However, when confronted with the idea that women leave natural science to fulfil goals that they perceive to be mismatched with the nature of academia, 57.5% supported this statement. In summary, the presented data provides as a comprehensive foundation to discuss **Hypothesis 1** and **Hypothesis 2** in the following **Section 5.1** and **Section 5.1**, respectively.*

Discussion 5

*Chapter 5 discusses the above presented results. To this end, single findings are evaluated in the grand scale and interpreted accordingly. The reader gains an in-depth understanding of how the obtained results contribute to either **Hypothesis 1** or **Hypothesis 2**. For the former, statistical evidence is used to critically review **Hypothesis 1** across different nations. For the latter, the diverse backgrounds of study participants stress the significance of the conducted study. To analyse **Hypothesis 2**, respective findings on study participants' future aspirations, individual perceptions as well as perceived reasons to stay or leave are contrasted to their respective beliefs on female underrepresentation in natural science / academia. Contrary assertions are critically highlighted and interpreted in a global picture. Finally, this section provides the foundation to draw a profound résumé in **Chapter 7**.*

5.1 Hypothesis 1: Gender Distribution in Natural Science

Hypothesis 1:   *Female underrepresentation predominately persists in senior positions in natural science / academia.*

This thesis aimed to unveil gender distributions in natural science / academia. To this end, recent statistics from Germany, the EU and U.S. were evaluated. The respective results were described in **Section** 4.1 and shall be discussed in respect to the postulated **Hypothesis 1** in the following section.

© The Author(s), under exclusive license to Springer Fachmedien Wiesbaden GmbH, part of Springer Nature 2023
A. Wolfram, *Extending the Complexity of the Leaky Pipeline Phenomenon in Natural Science*, BestMasters, https://doi.org/10.1007/978-3-658-43086-3_5

5.1.1   Gender Distribution in Natural Science – Germany

When analysing the presented data from **Section 4.1.1**, decline in female shares with increasing career stage could be observed. However, the decrease was not proportionally distributed across single career levels. In fact, near-gender parity was reached at M.Sc. graduate and PhD student level. Female shares stayed above 40% at PhD graduate and newly appointed junior professor levels. Then, a steep decline by more than 10 percentage points could be noted at higher levels (e.g., habilitation and W2 / W3 professorship). Thus, female underrepresentation does not exist in early career stages in Germany. In other words, enough women entered the career pipeline in natural science / academia. In fact, both women and men stay for a significant amount of time in the career path. For instance, at the level of PhD graduate, scientists have already completed a B.Sc. (minimum of 3 years), a M.Sc. (minimum of 2 years) and a PhD (minimum of 3 years). Before obtaining a junior professorship, most scientists tend to work as postdocs for several months to years. Thus, both men and women have dedicated more than 10 years of their lives to natural science / academia before a drastic decline in female shares can be seen. In a nutshell, female underrepresentation only persists in higher career levels. These findings validate **Hypothesis 1** for the academic research landscape in Germany.

5.1.2   Gender Distribution in Natural Science – Europe

At EU level, women accounted for the majority of bachelor's and master's students as well as graduates. In addition, gender parity was nearly reached at PhD student and graduate level. Thus, similar to the findings in Germany, both men and women were represented in early academic career levels. However, the reader should be reminded that the data sets refer to *typical academic career* without further exemplifications. Thus, comparability of the German and EU levels is somewhat limited. Nevertheless, on European level, near gender-balance was still reached at job entrance level (grade A) in academia. At the transition to more senior grade B positions, a first significant decline (7 percentage points) in women could be observed. At grade C, the most senior positions, female shares were drastically reduced to 26%. In conclusion, a stark underrepresentation could be observed in the most senior positions in academia (e.g., professorship). In short, whilst the career pipeline was over-proportionally filled with women, few reached top positions in academia.

The presented data clearly falsify the concept that there is a lack of female talent in academia; the contrary could be identified. The majority of academic talent was female. Thus, the EU gender distribution opens the question whether female talent is lost or actively hindered along the way to positions with decision-making power in academia. When evaluating gender distributions between 2015 and 2018, very little progress was made in higher positions. Seeing that gender distributions amongst B.Sc., M.Sc. and PhD students and graduates stayed identical, it seems peculiar that female share did not increase more significantly in top positions. Based on the amount of female talent in the career pipeline, one would expect a more significant increase over the span of three years.

In a nutshell, despite gender parity in early career levels, very limited progress was achieved in senior positions. Women stayed significantly underrepresented in higher academic positions which reinforces the validity of **Hypothesis 1**.

5.1.3 Gender Distribution in Natural Science – U.S.

In the U.S., the presented statistics revealed gender parity in B.Sc. STEM degrees. In M.Sc. STEM degrees, women made up 60% and were therefore overrepresented. At PhD level, gender distribution was nearly equal. When differentiating between life and physical science, women were overrepresented over all levels of higher education in life science. In physical science, the female share amounted to 40% in B.Sc. degrees. Then, a slight decline could be observed; at PhD level, women were represented with 36%.

In conclusion, women reached overall gender parity in STEM and dominated life sciences in higher education. Thus, in respect to **Hypothesis 1**, trends observed in Germany and the EU, could also be identified in the U.S.; female talent entered the career pipeline and female shares stayed high throughout higher education. Only in physical sciences, women were underrepresented at entrance levels, however, the respective shares stayed near constant. Interestingly, the Pew Research Center published overall shares of women in the STEM work force; Women had reached gender parity in STEM jobs between 2017–2019. In life and physical sciences, women accounted for 48% and 40% respectively. Please note that this data set included all STEM jobs (e.g., industry and academia). Surprisingly, female shares in life science were below the shares seen in higher education. Thus, it seems that women were lost at the transition between higher education and the job market. In contrast, women accounted for a higher percentage in the work force than at PhD level in physical jobs. Therefore, it seems that women with a background in physical science transitioned into the work environment without major losses. In fact, there is no *leaky pipeline phenomenon* detectable; women accounted for 40% of B.Sc. and workers in physical science. Thus, shares between degree entrance and work force can been seen as constant. In a nutshell, based on initial gender distributions, gender balance was kept. However, this was only true for the overall workforce. When comparing gender distributions in senior faculty positions and higher education, women are significantly underrepresented in STEM professorship positions. For instance, women only accounted for 23% of professors in 2021. Yet, in 2018, they represented 48% of PhDs in STEM. Thus, vertical gender segregation was clearly distinguishable in natural science / academia. In summary, female underrepresentation did not exist across the overall workforce but was evident in senior academic career levels in STEM. These findings emphasise the validity of **Hypothesis 1** in the U.S.

5.2 Hypothesis 2: Women in Natural Science – a Qualitative Analysis

Hypothesis 2:   *The goal congruity theory provides an incomplete explanation for female underrepresentation in natural science.*

To address **Hypothesis 2**, an international qualitative was conducted. To this end 26 women in science shared their future aspirations, their perceptions of the scientific community (see **Section 5.2.2**) as well as their perceived reasons to either stay in or leave natural science / academia (see **Section 5.2.3**). In addition, all of them were confronted with a statement that represents the *goal congruity theory*:

> *Women choose to leave natural science to fulfil goals that they perceive to be mismatched with the nature of academia.*

Finally, women shared their beliefs on female underrepresentation in science (see **Section 5.2.4**). Before discussing the respective findings, based on study participant characteristics, the general significance of the here presented qualitative study will be highlighted in **Section 5.2.1**.

5.2.1   Characterisation of Study Participants

Within natural science, the 26 study participants represented 17 vastly differing research fields (e.g., astronomy, computational biochemistry, neurobiology, organic chemistry). Thus, common trend identified in this study are most likely identifiable in the greater scientific community. In addition, women originated from 13 different countries across 5 continents. However, at the time of the interview, participants were primarily employed in North America, Europe and Australia and Oceania (i.e., New Zealand). Thus, it is most likely that common trends are associated with work affiliations in the western world. In fact, 25 out of 26 participants had worked either in North America or Europe in the past. In conclusion, the presented results hold most significance in these two continents. Nevertheless, during their respective academic paths, the interviewed women had worked in 79 countries and U.S. states. More than 90% had work experiences in two different nations / U.S. states. Therefore, the here presented data is the reflection of highly trained scientists with international work experiences. In conclusion, it seems more than probable that global trends identified in this study hold cross-national and cross-continental significance.

When analysing the personal background of study participants, it is notable that more than two thirds were married or in a long-term partnership. Thus, despite great labour mobility, the women managed to hold a partnership. However, few of them had children. In fact, only 3 women residing in natural science / academia stated to have children. Among them, one had a new-born and two had a toddler. The youngest of these women was 30, the oldest was 40 years old. Here, a global phenomenon can be observed. Women with higher education tend to have children later than the societal average. On average, women with a master's degree have their first child after they turn 30 (Livingston, 2015) in the U.S. When factoring in that all study participants also completed a PhD, it seems little surprising that the vast majority of them were childless. However, 10 / 21 desired children. Out of them, 8 were employed on short-term contracts. In particular, the required mobility at the postdoctoral research stage impeded with the desire to have children. In fact, all 8 stated to wait for a more permanent position.

5.2.2   Future Aspirations and Perception of Natural Science / Academia

The vast majority expressed the explicit desire to stay in academia. Out of 23 scientists that were actively employed within academia at the time of the interview,

19 were certain that they wanted to continue the academic path. This translates to a share of 82.6%. Considering that these women were 35.8 years on average, most of them had spent more than 15 years in their respective scientific discipline. This result provides the first indication that female scientists do not perceive a mismatch between their individual goals and the nature of academia. In the contrary, their major goal was to actively stay in academia. This is also underlined by their stated future aspirations. Out of all, 17 named a position (either *professorship, PI position* or *tenure track position*) as a major goal. At first, this seems surprising, as one could expect more science-orientated goals. However, when considering the desire for a senior position in combination with the average age, the civil status as well as the desire to have children, it becomes clear that women tend to desire positions to gain stability in their lives.

Other than position-oriented goals, the scientists listed science- and value-oriented goals. For the former, 9 women wanted to be *good scientists, mentors* and *teachers* which already implies the desire to fulfil communal goals within natural science / academia. For the later, 7 participants were desiring *to do things of value, NGO work* or *implementation-based work*. Here, for the first time, communal goals outside of natural science / academia were listed. In addition, 5 women expressed industry-based goals. In summary, only 7 out of 43 answers implied a desire to fulfil communal goals that were perceived as unaffordable in natural science / academia. This translates to a share of 16.3%. In conclusion, most study participants wanted to pursue their (communal) goals in academia which contradicts the *goal congruity theory*. Interestingly, when asked to describe the scientific community, study participants predominantly listed attributes associated with communal values such as *collaborative* (10 /26), *community* (4 / 26), *fostering* (2 / 26), *friendly* (2 / 26), *motivated* (2 / 26), *networking* (2 / 26), *sharing knowledge* (2 / 26), *supportive* (2 / 26), *collegial* (1 / 26), *communicative* (1 / 26), *connected* (1 / 26), *sense of wanting to help* (1 / 26). This aligns with the fact, that two thirds of all study participants have an overall positive perception of the scientific community.

In a nutshell, participants (to some extend unknowingly) illustrated their individual perceptions of natural science / academia by listing attributes for the scientific community. The interpretation of positive, neutral and negative attribute unveiled an overall positive as well as communal perception. These findings are in contrast to the *goal congruity theory* that proposes an inverse correlation between communal goal affordance and to STEM interest.

5.2.3   Reasons to Stay in / Leave Natural Science / Academia

To the answer why people (gender neutral) would stay in natural science / academia, study participants only listed 8 reasons. Among them, 3 accounted for communal values (e.g., *ability to mentor, ability to teach* and *teamwork, collaboration and exchange*). In fact, the perceived affordance of communal values within natural science / academia majorly contributed to the desire of female scientists to stay. The 5 other aspects can be clustered to individual goals and values (i.e., *pursuing research, freedom, love / passion of science, flexibility* and *personal satisfaction* can be seen as individualistic goals that serve oneself).

In contrast, when evaluating clustered reasons why scientists (gender neutral) leave natural science / academia, only one aspect somewhat indicated towards communal values; 5 women stated [toxic] work environments as reason to leave. In fact, all other aspects were centred around the individuum; in first place, personal and professional instability is perceived to drive people out of science. In second place, both mental health and workload are cited. By a distance, [low] salaries and individual choices are held responsible for people opting out.

Most strikingly, financial concerns are weaved throughout all aspects. For instance, as a response to lack of funding (both for wage(s) and consumables), scientists will search for grants. After choosing appropriate grants, scientists are most likely to write several applications to maximise success chances. Depending on the funding agency, it will take between weeks to months to write a single grant application. Regularly, scientists will work in the lab, supervise students, mentor mentees and teach during the day and work on their applications in their off-time. The workload in combination with financial insecurities, pressure to obtain funding and little work-life-balance negatively affects mental wellbeing over time. In conclusion, people either leave out of frustration or for higher salaries or, in the worst-case scenario, because of severe mental health issues. Thus, financial insecurity is the major problem which may even force people out of science that rather wanted to stay. Going back to **Section 5.2.2**, the lack of financial instability can also be seen as a strong motivator to aspire a more permanent and better paid as well as funded position within natural science / academia. During the one-on-one interviews, many of the study participants described to be in a short-term position without knowing what would come next. Some applied for grants whilst also applying outside of science, others lived on credit cards in between contracts. In extreme cases, women had worked on 3-months contracts for years. The precarious financial situation of people is a global problem across all disciplines in natural science.

In addition to harming the mental well-being of effected scientists, financial insecurity may also negatively impact the research output and therefore the whole lab group; PIs profit from highly performing short-term junior staff (PhD students and postdocs) as they drive scientific progress within a relative short time. However, there is only so much time within one day. If people do not have a (semi-) secure workplace, time will be dedicated to finding the next scholarship, grant or position. In addition, time will be lost worrying about the future. This is time that scientists cannot perform at their best and which negatively impacts their personal and group output.

5.2.4   *Goal Congruity Theory* and Female Underrepresentation in Science / Academia

Surprisingly, when confronted with the idea that women leave natural science to fulfil goals that they perceive to be mismatched with the nature of academia, 57.5% of study participants supported the *goal congruity theory*. This seems to be in stark contrast to the previously mentioned reasons why scientists leave natural science / academia (i.e., *uncertainty*, *mental wellbeing* and *workload*). In addition, women shared their beliefs on female underrepresentation in science; Stereotypes and gender roles were most commonly cited (61.5%). When listening to the single women, it became evident that women found themselves and others (including men) struggling in natural science / academia (e.g., due to financial and job instability). Being asked to think about other women, many jumped to the conclusion that women who wanted to have children in science, had to face all the hardships everyone was facing, but in addition would also face the stereotypes and gender roles associated with motherhood. In line, many expressed sympathies for female scientists who chose to leave natural science to fulfil other goals / roles including motherhood.

Interestingly, those scientists who disagreed with the *goal congruity theory* (6 / 26), used a different narrative; due to lack of funding, job stability, lack of childcare and long working hours as well as societal pressure, women (with children and / or family) may leave science, not because they have children and / or family but because of appalling work conditions that are unsustainable for people with family commitment.

This narrative focuses on underlying work conditions as a major obstacle to gender parity in science. Thus, the criticism points towards the academic system held upright by funding agencies, universities, institutes and the people working in them. Here, change could be provoked by enforcing certain laws or policies (e.g., funding policies regulating a minimum contract-length of 3 years). In contrast, in a narrative where women chose to leave science to fulfil their individual goals, decision makers are off the hook and little change will be evoked.

*In **Chapter 5** previously presented data were interpreted to address either **Hypothesis 1** or **Hypothesis 2**. In the former, based on statistical evidence from German, EU and U.S. female underrepresentation could be identified as a global trend. In line with **Hypothesis 1** (near-)gender parity could be observed in early career stages. Along the secondary and tertiary education, female shares stayed approximately constant. Drastic changes were seen in more senior positions. In single most senior positions, women were significantly underrepresented. In the latter, participants' future aspirations, individual perceptions, perceived reasons to stay or leave were critically contrasted to the goal congruity theory. Remarkably, study participants illustrated an overall positive as well as communal perception of natural science / academia. In fact, the perceived affordance of communal values within science majorly contributed to the desire of female researcher to stay in the academic path. These findings contradict the literature-established goal congruity theory. However, when confronted with the idea that women leave natural science to fulfil goals they perceived to be mismatched with the nature of academia, 57.5% of study participants supported the goal congruity theory. Analysing individual explanations, sympathy towards women having to deal with the instable nature of academia as well as being exposed to stereotypes associated with working mothers were the common motives why study participants agreed. After having discussed all presented data, the respective significance of this thesis shall be highlighted in **Chapter 6**. Afterwards, a final résumé will be drawn in **Chapter 7**.*

Limitations

6

*After having discussed all presented data, this section will shed light on respective explanatory power in the grand scale. To this end, meaningfulness and limitations of collected, analysed and discussed data shall be highlighted. In particular, limitations associated with the cross-comparability of single data sets contributing to **Hypothesis 1** are illustrated. For **Hypothesis 2**, the significance of respective findings is linked to study size, study participants and effects such as social-desirability bias.*

6.1   Hypothesis 1: Gender Distribution in Natural Science

Whilst all presented data sets contributed to **Hypothesis 1**, the respective cross-comparability is somewhat limited. In fact, the included statistics analysed different stakeholder groups. For Germany, statistics included people from natural science and mathematics. In the EU, numbers from *typical academic careers* and in the US, statistics for STEM, life and physical science were presented. Hence, whilst conclusive in itself, gender distributions in Germany, the EU and U.S. describe drastically differing academic /scientific groups. In particular, data from the EU holds limited explanatory power due to imprecise characterisation. The term *typical academic career* holds great room for interpretation and may hold different meaning in each of the EU member countries. Nevertheless, all

Supplementary Information The online version contains supplementary material available at https://doi.org/10.1007/978-3-658-43086-3_6.

findings describe a global trend; vertical gender segregation in natural science / academia persists in Germany, the EU and U.S.

Further, it should be mentioned that presented data referred to a timespan between 2015–2021. In fact, it was rather difficult to obtain comprehensive numbers for most recent gender distributions; recent cross-national data were only available for 2019. Thus, the majority of statistics represent respective gender distributions in 2018 and 2019. In consequence, potential impacts of the COVID-19 pandemic are not accounted for.

In addition, most statistics compare gender distributions across career levels within the same year or period. This may limit the interpretability. To reach higher positions in natural science / academia, young talent needs to complete higher education including a PhD. In addition, it is common to work as a postdoc to gain more lab experience before transitioning to a habilitation or a junior professorship. Consequently, depending on the individual career path, many years pass before an up-and-coming scientist may obtain a professorship. In fact, based on statical analysis in 2020, researchers were 41.7 years (W2) and 43.2 years (W3) on average when obtaining their professorship in Germany (Krabel et al., 2021, p. 91). Hence, both men and women dedicated approximately 20 years to natural science / academia before transitioning to a senior position. In this context, it would be more appropriate to compare respective gender distributions over time (i.e., female shares of higher education levels in 2000, PhD graduation in 2010 and newly appointed professorships in 2020).

In this context, Charlesworth and Banaji reviewed female shares across higher education in the U.S. between 2000 and 2015 (2019). In Physical sciences, gender distributions in bachelor's and master's degrees stayed overall constant. At PhD level, a constant but moderate increase in women was detected. However, in the same timespan, female shares at associate level drastically declined from above 50% (2000) to 40% (2015). In biology, female shares across B.Sc., M.Sc. and associate level stayed overall constant. An increase from below to above 50% could be seen for PhDs. In summary, the respective findings indicate towards rather constant gender distributions across higher educating in natural science. In other words, female overrepresentation in biology and relatively high female representation in physical science (approx. 40%) has been observed in the U.S. over a 20-year period. These exemplary findings further underline **Hypothesis 1**; female underrepresentation persists in senior position despite constant and high female shares in early career levels.

6.2 Hypothesis 2: Women in Natural Science – a Qualitative Analysis

To address **Hypothesis 2**, an international qualitative study was set up. Here, already in the recruitment of study participants, a bias can be identified. Within the three strategies pursued, a clear difference in response of study participants can be identified:

A) Contacting postdocs/ junior group leaders within the scientific network of the author (**5 / 68** invited women, **4 / 26** study participants),
B) Contacting senior staff within the scientific network of the author to access their network and contacts (**15 / 68** invited women, **8 / 26** study participants),
C) Searching *Gage*—the world's largest directory of women and gender diverse folks in science (Gage. Discover Brilliance)—for female scientists from around the globe to invite them and access their network and contacts (**48 / 68** invited women, **14 / 26** study participants).

From those who received an invitation email, 80% of direct contacts, 53% of indirect contacts and 30% of strangers responded and agreed to participate. In total, 46% of study participants were directly or indirectly associated with the investigator. This could limit the global significance of the presented trends, as nearly half of all participants belong to a global but interconnected bubble. The other 14 participants that had no prior ties to the investigator and were mostly registered on *Gage*—the world's largest directory of women and gender diverse folks in science. This could imply that these women were overly aware of gender imbalances in natural sciences as they actively signed up on a platform to promote female visibility in science. The total number of participants accounted to 26. Whilst these women had a background in 17 vastly differing research fields, originated from 13 different countries across 5 continents and had international work experiences in 79 countries, statistical significance could be enhanced by including more study participants preferably from continents that were underrepresented in this study (e.g., Africa and South America). However, the time frame associated with this master's project was limited to 16 weeks and thus prevented in inclusion of further participants. It should be mentioned that it would be of great value to repeat this study setup with men in natural science / academia. The respective findings could be contrasted to the here presented results. In fact, this would allow to analyse gender-depended response bias. In line with response bias, social-desirability bias—the tendency of study participants to answer questions in a manner that they perceived to be favoured—needs to be highlighted.

Whilst the semi-structured interview guide loosely based on Cornelia Helfferich was designed with very broad questions to allow participants to share what they felt appropriate, all participants received an invitation email (see **EMS**, p. 2) in which the investigator outlined the general topic as *hard and soft barriers Women in Science encounter*. Thus, participants were aware of the direction of the interview. All of the above describe limitations should be factored in when evaluating the overall significance of the results of this qualitative study. However, due to the diversity of study participants, on can assume that the identified trends reflect trends in natural science / academia. In addition, the global nature of science and the scientific community also indicates towards greater significance.

*In **Chapter 6**, limitations in data validity and expressiveness associated with **Hypothesis 1** and **Hypothesis 2** were elaborated. In the former, cross-comparability, insufficient data description and original timespan data collection were named as major contains. Despite these limitations, the common trend—female underrepresentation in senior position despite near-constant and high female shares in early career levels—stayed valid. For the latter, study size, underrepresentation of women from Africa and South America, associations between study participants and lack of male participant were elaborated as limitations. In addition, effects such as social-desirability bias minimised validity. Nevertheless, due to the global nature of science and the diverse backgrounds of study participants, identified trends can still be seen as significant.*

Conclusion and Greater Benefit

7

*In **Section 7**, the final résumé is drawn. In attempt to answer the two initial questions 1) Does the leaky pipeline phenomenon continue to exist in natural science / academia and if so, where are major leaks in the pipeline? and 2) Do female scientists leave natural science / academia to afford other goals? the author endorses both **Hypothesis 1** and **Hypothesis 2**. In addition, key learnings and their impact on women in science are revealed. Finally, the greater benefit of the here presented thesis is illustrated.*

7.1   Final remarks

"You can't be what you can't see." (Sally Ride, astronaut and physicist)

Dr Sally K. Ride—the first American woman in space—once said that "young girls need to see role models in whatever careers they may choose, just so they can picture themselves doing those jobs someday" (Beard, 2012). This is also true for natural science / academia; as long as senior positions will be associated with men, gender disparities will persist.

When recapitulating the findings of the here presented thesis, the analysis of statistical evidence from Germany, EU and U.S. revealed a clear trend: female underrepresentation in senior positions despite near-constant and high female shares in early career levels.

As hypothesised, female underrepresentation predominately persists in senior positions in natural science / academia. Across the secondary and tertiary education, female shares are approximately constant. Drastic changes in gender distributions are predominantly seen in more senior positions. In most senior positions, women are significantly underrepresented. Seeing that this trend is true for Germany, the EU and the U.S., is seems unlikely that time alone will achieve gender parity in decision-making positions.

Nevertheless, the academic pipeline is full of young talent—both men and women. In fact, the presented data revealed that both genders are retained within the pipeline. Leaks are rarely seen along the secondary and tertiary education. In life sciences, women are and have been overrepresented in early career levels since the 2000s. In the U.S., gender parity has been reached in the STEM work force. However, to increase female shares in senior positions, additional measures may be needed to be undertaken to reduce the time required to reach gender parity (e.g., female quota). In short, these findings imply that women do not opt out of STEM in early stages of the career. This serves as a preliminary indication that the *goal congruity theory* might not apply to women working in STEM. In fact, findings associated with **Hypothesis 2** strongly contradict the *goal congruity theory*; study participants predominantly used communal-oriented attributes to describe the scientific community. The women illustrated an overall positive as well as communal perception of natural science / academia. In addition, many study participants specifically named communal goals such as collaborating with others, mentoring and teaching students as reasons why they desired to stay in academia. In a nutshell, the perceived affordance of communal values within science significantly contributed to the desire of female researcher to stay in natural science / academia.

On the big scale, two key learnings can be deduced from the here presented study:

1. **On average, female scientists desire to stay in natural science as this allows them to pursue their valued goals** (e.g., pursuing research, mentoring and personal and scientific freedom). However, due to appalling funding, some cannot afford to stay in natural science / academia.

2. **Women disproportionately suffer a non-trivial conflict of timing in natural sciences / academia.** To continue with the academic path, postdocs are expected to move laboratories/ research groups. This requires mobility and often entails shifting countries or even continents during the mid 30s of young scientists. This is the same time that women may want to have children. In contrast to men, women are more effected by the biological clock. Thus, they have to face a timing problem which is not easily resolved.

Whilst the second aspect can only be partially addressed (e.g., providing 3-year postdoc contracts, day care facilities and supporting transitions into positions after contract termination), the first aspect can be directly addressed by decision makers and funding agencies and change is urgently required. In fact, already today, science is facing a global labour shortage in postdoctoral researchers. Not because there are none, but because scientists are discontent with working for below-average compensations despite being highly trained.

7.2   Greater Benefit

The greater benefits of this project can be described on three levels:

A)   Micro level: Individual benefit for the study participants

The study enabled female scientists to openly share their respective experiences and motives to either stay in or leave academia. As the study conductor acted independently from their respective universities / institutes, the fear of information spillage to officials was minimised. Thus, the study provided women with an opportunity to share details they had not shared before. Knowing that the data would be anonymised and statistically analysed, female scientists felt that they were heard without risking identity exposure.

B)   Meso level: Benefit for the universities / research institutes

For universities / research institutes, the here presented study can be seen as an educative experience-based analysis of the current situation. Universities / research institutes may exploit the study results as a foundation to deduce actions to counteract identified underlying barriers und thus, promote gender equality in natural science / academia.

C)   Marco level: Benefit for the scientific community

Currently, academia is suffering a labour shortage in postdoctoral researchers. To address this, the scientific community needs to understand why young scientists choose to leave the academic path. Thus, understanding the underlying reasons why women leave (or even feel they have to leave) is essential to address the labour shortage in natural science.

All-in-all, the qualitative study provides an experience-based foundation to 1) discuss and 2) rationally address the *leaky pipeline phenomenon* in natural science / academia. In fact, this work could be exploited by institutes and universities around the globe to conduct similar qualitative studies with local staff to identify institute-dependent barriers to equality. Ultimately, this would allow institutes around the globe to derive specific institute-adapted measures to promote equal opportunity in natural science.

In the end, the author of the here presented thesis unveils two simple but impactful learnings: 1) The majority of female researchers desires to stay in natural science as this allows them to pursue their valued goals, however, due to appalling funding, may not be able stay, and 2) Women disproportionately suffer a non-trivial conflict of timing in natural sciences / academia. Whilst the second can only be partially addressed, the first holds great potential of systemic changes in funding policies.

References

Auschra, C., Bartosch, J., & Lohmeyer, N. (2022). Differences in female representation in leading management and organization journals: Establishing a benchmark. *Research Policy, 51*(3), 104410. https://doi.org/10.1016/J.RESPOL.2021.104410

Ayub, M., Aamir Khan, R., & Khushnood, M. (2019). Glass Ceiling or Personal Barriers: A Study of Underrepresentation of Women in Senior Management. *Global Social Sciences Review (GSSR), IV*. https://doi.org/10.31703/gssr.2019(IV-IV).17

Babic, A., & Hansez, I. (2021). The Glass Ceiling for Women Managers: Antecedents and Consequences for Work-Family Interface and Well-Being at Work. *Frontiers in Psychology, 12*, 677. https://doi.org/10.3389/FPSYG.2021.618250

Beard, A. (2012, September). Sally Ride. *Harvard Business Review*. https://hbr.org/2012/09/sally-ride

Beardslee, D. C., & O'Dowd, D. D. (1961). The college-student image of the scientist. *Science, 133*(3457), 997–1001. https://doi.org/10.1126/SCIENCE.133.3457.997

Benard, S., Paik, I., & Correll, S. J. (2007). Cognitive Bias and the Motherhood Penalty. *Hastings Law Journal, 59*. https://repository.uclawsf.edu/hastings_law_journal/vol59/iss6/3

Blickenstaff, J. C. (2006). Women and science careers: leaky pipeline or gender filter? *Gender and Education, 17*(4), 369–386. https://doi.org/10.1080/09540250500145072

Brooks, J., & Della Sala, S. (2009). Re-addressing gender bias in Cortex publications. *Cortex; a Journal Devoted to the Study of the Nervous System and Behavior, 45*(10), 1126–1137. https://doi.org/10.1016/J.CORTEX.2009.04.004

Brown, E. R., Thoman, D. B., Smith, J. L., & Diekman, A. B. (2015). Closing the Communal Gap: The Importance of Communal Affordances in Science Career Motivation. *Journal of Applied Social Psychology, 45*(12), 662. https://doi.org/10.1111/JASP.12327

Ceci, S. J., Ginther, D. K., Kahn, S., & Williams, W. M. (2014). Women in Academic Science: A Changing Landscape. *Psychological Science in the Public Interest : A Journal of the American Psychological Society, 15*(3), 75–141. https://doi.org/10.1177/1529100614541236

Ceci, S. J., & Williams, W. M. (2011). Understanding current causes of women's underrepresentation in science. *Proceedings of the National Academy of Sciences of the United States of America, 108*(8), 3157–3162. https://doi.org/10.1073/PNAS.1014871108

Ceci, S. J., Williams, W. M., & Barnett, S. M. (2009). Women's underrepresentation in science: sociocultural and biological considerations. *Psychological Bulletin, 135*(2), 218–261. https://doi.org/10.1037/A0014412

Charlesworth, T. E. S., & Banaji, M. R. (2019). Gender in Science, Technology, Engineering, and Mathematics: Issues, Causes, Solutions. *Journal of Neuroscience, 39*(37), 7228–7243. https://doi.org/10.1523/JNEUROSCI.0475-18.2019

Correll, S. J. (2016). Constraints into Preferences: Gender, Status, and Emerging Career Aspirations. *American Sociological Association, 69*(1), 93–113. https://doi.org/10.1177/000312240406900106

Correll, S. J., Benard, S., & Paik, I. (2007). Getting a Job: Is There a Motherhood Penalty? *American Journal of Sociology, 112*(5), 1297–1338. https://doi.org/10.1086/511799

Cuddy, A. J. C., Fiske, S. T., & Glick, P. (2004). When Professionals Become Mothers, Warmth Doesn't Cut the Ice. *Journal of Social Issues, 60*(4), 701–718. https://doi.org/10.1111/J.0022-4537.2004.00381.X

Diekman, A. B., Brown, E. R., Johnston, A. M., & Clark, E. K. (2010). Seeking Congruity Between Goals and Roles: A New Look at Why Women Opt Out of Science, Technology, Engineering, and Mathematics Careers. *Psychological Science, 21*(8), 1051–1057. https://doi.org/10.1177/0956797610377342

Diekman, A. B., Clark, E. K., Johnston, A. M., Brown, E. R., & Steinberg, M. (2011). Malleability in communal goals and beliefs influences attraction to stem careers: evidence for a goal congruity perspective. *Journal of Personality and Social Psychology, 101*(5), 902–918. https://doi.org/10.1037/A0025199

Diekman, A. B., Joshi, M. P., & Benson-Greenwald, T. M. (2020). Goal congruity theory: Navigating the social structure to fulfill goals. *Advances in Experimental Social Psychology, 62*, 189–244. https://doi.org/10.1016/BS.AESP.2020.04.003

Diekman, A. B., Steinberg, M., Brown, E. R., Belanger, A. L., & Clark, E. K. (2017). A Goal Congruity Model of Role Entry, Engagement, and Exit: Understanding Communal Goal Processes in STEM Gender Gaps. *Personality and Social Psychology Review, 21*(2), 142–175. https://doi.org/10.1177/1088868316642141

European Commission, & Innovation Directorate-General for Research. (2021). *She figures 2021: gender in research and innovation: statistics and indicators*. Publications Office. https://doi.org/10.2777/06090

European Women on Boards. (2021). *GENDER DIVERSITY INDEX OF WOMEN ON BOARDS AND IN CORPORATE LEADERSHIP*. https://europeanwomenonboards.eu/wp-content/uploads/2022/01/2021-Gender-Diversity-Index.pdf

eurostat. (2021, March 5). *Women remain outnumbered in management - Produkte Eurostat Aktuell - Eurostat*. https://ec.europa.eu/eurostat/de/web/products-eurostat-news/-/edn-20210305-2

Fernandez-Mateo, I., & Fernandez, R. M. (2016). Bending the Pipeline? Executive Search and Gender Inequality in Hiring for Top Management Jobs. *Management Science, 62*(12), 3636–3655. https://doi.org/10.1287/MNSC.2015.2315

Fry, R., Kennedy, B., & Funk, C. (2021). *STEM Jobs See Uneven Progress in Increasing Gender, Racial and Ethnic Diversity | Pew Research Center*. https://www.pewresearch.org/science/2021/04/01/stem-jobs-see-uneven-progress-in-increasing-gender-racial-and-ethnic-diversity/

Fuesting, M. A., & Diekman, A. B. (2017). Not By Success Alone: Role Models Provide Pathways to Communal Opportunities in STEM. *Personality & Social Psychology Bulletin, 43*(2), 163–176. https://doi.org/10.1177/0146167216678857

gage. *Discover Brilliance.* (n.d.). Retrieved October 4, 2022, from https://gage.500womenscientists.org/

Gewin, V. (2018). Why diversity helps to produce stronger research. *Nature.* https://doi.org/10.1038/D41586-018-07415-9

Goulden, M., Mason, M. A., & Frasch, K. (2011). Keeping Women in the Science Pipeline. *AAPSS, 638*(1), 141–162. https://doi.org/10.1177/0002716211416925

Halpern, D. F., Benbow, C. P., Geary, D. C., Gur, R. C., Hyde, J. S., & Gernsbacher, M. A. (2007). The science of sex differences in science and mathematics. *Psychological Science in the Public Interest, Supplement, 8*(1), 1–51. doi: https://doi.org/10.1111/j.1529-1006.2007.00032.x

Halpern, D. F., Straight, C. A., & Stephenson, C. L. (2011). Beliefs About Cognitive Gender Differences: Accurate for Direction, Underestimated for Size. *Sex Roles, 64*(5–6), 336–347. https://doi.org/10.1007/s11199-010-9891-2

Helfferich, C. (2011). Die Qualität qualitativer Daten. In *Die Qualität qualitativer Daten* (Issue 4). VS Verlag für Sozialwissenschaften. https://doi.org/10.1007/978-3-531-92076-4

Hodapp, T., & Brown, E. (2018). Making physics more inclusive. *Nature 2021 557:7707, 557*(7707), 629–632. https://doi.org/10.1038/d41586-018-05260-4

Holman, L., Stuart-Fox, D., & Hauser, C. E. (2018). The gender gap in science: How long until women are equally represented? *PLoS Biology, 16*(4). https://doi.org/10.1371/JOURNAL.PBIO.2004956

Hosek, S. D., Cox, A. G., Ghosh-Dastidar, B., Kofner, A., Ramphal, N., Scott, J., & Berry, S. H. (2005). *Gender Differences in Major Federal External Grant Programs.* http://www.rand.org/

Howe-Walsh, L., & Turnbull, S. (2014). Barriers to women leaders in academia: tales from science and technology. *Studies in Higher Education, 41*(3), 415–428. https://doi.org/10.1080/03075079.2014.929102

Huang, J., Gates, A. J., Sinatra, R., & Barabási, A. L. (2020). Historical comparison of gender inequality in scientific careers across countries and disciplines. *Proceedings of the National Academy of Sciences of the United States of America, 117*(9), 4609–4616. https://doi.org/10.1073/PNAS.1914221117

Huguet, P., & Régner, I. (2009). Counter-stereotypic beliefs in math do not protect school girls from stereotype threat. *Journal of Experimental Social Psychology, 45*(4), 1024–1027. https://doi.org/10.1016/J.JESP.2009.04.029

Hyde, J. S. (2014). Gender similarities and differences. *Annual Review of Psychology, 65*, 373–398. https://doi.org/10.1146/ANNUREV-PSYCH-010213-115057

Hyde, J. S. (2016). Sex and cognition: gender and cognitive functions. *Current Opinion in Neurobiology, 38*, 53–56. https://doi.org/10.1016/J.CONB.2016.02.007

Hyde, J. S., Lindberg, S. M., Linn, M. C., Ellis, A. B., & Williams, C. C. (2008). Diversity: Gender similarities characterize math performance. *Science, 321*(5888), 494–495. https://doi.org/10.1126/SCIENCE.1160364

Hyde, J. S., & Mertz, J. E. (2009). Gender, culture, and mathematics performance. *Proceedings of the National Academy of Sciences, 106*(22), 8801–8807. https://doi.org/ 10.1073/PNAS.0901265106

International Labour Office. (2016). *Women at Work: Trends 2016*. www.ilo.org/publns

Keller, E. F., & Scharff-Goldhaber, G. (1998). Reflections on Gender and Science. *American Journal of Physics, 55*(3), 284. https://doi.org/10.1119/1.15186

Kozlowski, D., Lariviere, V., Sugimoto, C. R., & Monroe-White, T. (2022). Intersectional inequalities in science. *Proceedings of the National Academy of Sciences of the United States of America, 119*(2), e2113067119. https://doi.org/10.1073/PNAS.2113067119

Krabel, S., Hauss, K., Shajek, A., Staneva, M., & Schmid, S. (2021). *Bundesbericht Wissenschaftlicher Nachwuchs 2021*. https://doi.org/10.3278/6004603aw

Kurdi, B., Seitchik, A. E., Axt, J. R., Carroll, T. J., Karapetyan, A., Kaushik, N., Tomezsko, D., Greenwald, A. G., & Banaji, M. R. (2019). Relationship between the implicit association test and intergroup behavior: A meta-analysis. *American Psychologist, 74*(5), 569–586. https://doi.org/10.1037/AMP0000364

Livingston, G. (2015, January 15). *For most highly educated women, motherhood doesn't start until the 30s | Pew Research Center*. https://www.pewresearch.org/fact-tank/2015/ 01/15/for-most-highly-educated-women-motherhood-doesnt-start-until-the-30s/

Makarova, E., Aeschlimann, B., & Herzog, W. (2019). The Gender Gap in STEM Fields: The Impact of the Gender Stereotype of Math and Science on Secondary Students' Career Aspirations. *Frontiers in Education, 4*, 60. https://doi.org/10.3389/FEDUC.2019.00060

Makarova, E., & Herzog, W. (2015). Trapped in the gender stereotype? the image of science among secondary school students and teachers. *Equality, Diversity and Inclusion, 34*(2), 106–123. https://doi.org/10.1108/EDI-11-2013-0097

Marinak, B. A., & Gambrell, L. B. (2010). Reading Motivation: Exploring the Elementary Gender Gap. *Literacy Research and Instruction, 49*(2), 129–141. https://doi.org/10.1080/ 19388070902803795

McKinnon, M., & O'Connell, C. (2020). Perceptions of stereotypes applied to women who publicly communicate their STEM work. *Humanities and Social Sciences Communications 2020 7:1, 7*(1), 1–8. https://doi.org/10.1057/s41599-020-00654-0

Meho, L. I. (2021). The gender gap in highly prestigious international research awards, 2001–2020. *Quantitative Science Studies, 2*(3), 976–989. https://doi.org/10.1162/QSS_ A_00148

Mehta, N. (2022, April 4). *Faculty, students reflect on STEM faculty gender gap*. https:// www.browndailyherald.com/article/2022/04/faculty-students-reflect-on-stem-faculty-gender-gap

Nosek, B. A., & Smyth, F. L. (2007). A multitrait-multimethod validation of the implicit association test: Implicit and explicit attitudes are related but distinct constructs. *Experimental Psychology, 54*(1), 14–29. https://doi.org/10.1027/1618-3169.54.1.14

Nosek, B. A., & Smyth, F. L. (2011). Implicit Social Cognitions Predict Sex Differences in Math Engagement and Achievement: *American Educational Research Journal, 48*(5), 1125–1156. https://doi.org/10.3102/0002831211410683

Oakley, J. G. (2000). Gender-based Barriers to Senior Management Positions: Understanding the Scarcity of Female CEOs. *Journal of Business Ethics 2000 27:4, 27*(4), 321–334. https://doi.org/10.1023/A:1006226129868

Office of Federal Operations. (2019). *Annual Report on the Workforce, US Equal Employment Opportunity Commission, Special Topic: Women in STEM*. https://www.eeoc.gov/special-topics-annual-report-women-stem

Oliveira, D. F. M., Ma, Y., Woodruff, T. K., & Uzzi, B. (2019). Comparison of National Institutes of Health Grant Amounts to First-Time Male and Female Principal Investigators. *JAMA, 321*(9), 898–900. https://doi.org/10.1001/JAMA.2018.21944

Pohlhaus, J. R., Jiang, H., Wagner, R. M., Schaffer, W. T., & Pinn, V. W. (2011). Sex Differences in Application, Success, and Funding Rates for NIH Extramural Programs. *Academic Medicine, 86*(6), 759. https://doi.org/10.1097/ACM.0B013E31821836FF

Prager, K. (2021). Seek diversity to solve complexity. *Nature*. https://doi.org/10.1038/D41586-021-01832-Z

Régner, I., Thinus-Blanc, C., Netter, A., Schmader, T., & Huguet, P. (2019). Committees with implicit biases promote fewer women when they do not believe gender bias exists. *Nature Human Behaviour, 3*(11), 1171–1179. https://doi.org/10.1038/S41562-019-0686-3

Reilly, D., Neumann, D. L., & Andrews, G. (2019). Gender differences in reading and writing achievement: Evidence from the National Assessment of Educational Progress (NAEP). *The American Psychologist, 74*(4), 445–458. https://doi.org/10.1037/AMP0000356

Resmini, M. (2016). The 'Leaky Pipeline'. *Chemistry – A European Journal, 22*(11), 3533–3534. https://doi.org/10.1002/CHEM.201600292

Reuben, E., Sapienza, P., & Zingales, L. (2014). How stereotypes impair women's careers in science. *Proceedings of the National Academy of Sciences of the United States of America, 111*(12), 4403–4408. https://doi.org/10.1073/PNAS.1314788111

Ross, M. B., Glennon, B. M., Murciano-Goroff, R., Berkes, E. G., Weinberg, B. A., & Lane, J. I. (2022). Women are credited less in science than men. *Nature 2022 608:7921, 608*(7921), 135–145. https://doi.org/10.1038/s41586-022-04966-w

Schinske, J., Cardenas, M., & Kaliangara, J. (2015). Uncovering scientist Stereotypes and their relationships with student race and student success in a diverse, community college setting. *CBE Life Sciences Education, 14*(3). https://doi.org/10.1187/CBE.14-12-0231

Seo, G., Huang, W., & Han, S. H. C. (2017). Conceptual Review of Underrepresentation of Women in Senior Leadership Positions From a Perspective of Gendered Social Status in the Workplace. *Human Resource Development Review, 16*(1), 35–59. https://doi.org/10.1177/1534484317690063

Statistisches Bundesamt (Destatis). (2022). *Bildung und Kultur, Personal an Hochschulen 2021*. https://www.destatis.de/DE/Themen/Gesellschaft-Umwelt/Bildung-Forschung-Kultur/Hochschulen/Publikationen/Downloads-Hochschulen/personal-hochschulen-211 0440217004.pdf;jsessionid=BFDFDA3CF23646C0887BD4459D4CB1A1.live722?__ blob=publicationFile

Sterling, A. D., Thompson, M. E., Wang, S., Kusimo, A., Gilmartin, S., & Sheppard, S. (2020). The confidence gap predicts the gender pay gap among STEM graduates. *Proceedings of the National Academy of Sciences of the United States of America, 117*(48), 30303–30308. https://doi.org/10.1073/PNAS.2010269117

Tannenbaum, C., Ellis, R. P., Eyssel, F., Zou, J., & Schiebinger, L. (2019). Sex and gender analysis improves science and engineering. *Nature, 575*(7781), 137–146. https://doi.org/10.1038/S41586-019-1657-6

The ten leading countries in natural-sciences research. (2020). *Nature*. https://doi.org/10.1038/D41586-020-01231-W

Voyer, D. (2011). Sex differences in dichotic listening. *Brain and Cognition*, *76*(2), 245–255. https://doi.org/10.1016/J.BANDC.2011.02.001

Voyer, D., Voyer, S., & Bryden, M. P. (1995). Magnitude of sex differences in spatial abilities: a meta-analysis and consideration of critical variables. *Psychological Bulletin*, *117*(2), 250–270. https://doi.org/10.1037/0033-2909.117.2.250

Voyer, D., Voyer, S. D., & Saint-Aubin, J. (2017). Sex differences in visual-spatial working memory: A meta-analysis. *Psychonomic Bulletin & Review*, *24*(2), 307–334. https://doi.org/10.3758/S13423-016-1085-7

Waisbren, S. E., Bowles, H., Hasan, T., Zou, K. H., Emans, S. J., Goldberg, C., Gould, S., Levine, D., Lieberman, E., Loeken, M., Longtine, J., Nadelson, C., Patenaude, A. F., Quinn, D., Randolph, A. G., Solet, J. M., Ullrich, N., Walensky, R., Weitzman, P., & Christou, H. (2008). Gender differences in research grant applications and funding outcomes for medical school faculty. *Journal of Women's Health (2002)*, *17*(2), 207–214. https://doi.org/10.1089/JWH.2007.0412

Watson, C. (2021). Women less likely to win major research awards. *Nature*. https://doi.org/10.1038/D41586-021-02497-4

Weisgram, E. S., & Bigler, R. S. (2006). Girls and science careers: The role of altruistic values and attitudes about scientific tasks. *Journal of Applied Developmental Psychology*, *27*(4), 326–348. https://doi.org/10.1016/J.APPDEV.2006.04.004

Weisgram, E. S., Dinella, L. M., & Fulcher, M. (2011). The Role of Masculinity/Femininity, Values, and Occupational Value Affordances in Shaping Young Men's and Women's Occupational Choices. *Sex Roles*, *65*(3), 243–258. https://doi.org/10.1007/S11199-011-9998-0

West, J. D., Jacquet, J., King, M. M., Correll, S. J., & Bergstrom, C. T. (2013). The role of gender in scholarly authorship. *PloS One*, *8*(7). https://doi.org/10.1371/JOURNAL.PONE.0066212

Witzel, A. (1982). *Verfahren der qualitativen Sozialforschung: Überblick und Alternativen.* Campus Verlag.

Yang, Y., Tian, T. Y., Woodruff, T. K., Jones, B. F., & Uzzi, B. (2022). Gender-diverse teams produce more novel and higher-impact scientific ideas. *Proceedings of the National Academy of Sciences of the United States of America*, *119*(36), e2200841119. https://doi.org/10.1073/PNAS.2200841119

Ysseldyk, R., Greenaway, K. H., Hassinger, E., Zutrauen, S., Lintz, J., Bhatia, M. P., Frye, M., Starkenburg, E., & Tai, V. (2019). A leak in the academic pipeline: Identity and health among postdoctoral women. *Frontiers in Psychology*, *10*(JUN), 1297. https://doi.org/10.3389/FPSYG.2019.01297

The manufacturer's authorised representative in the EU is Springer
Nature Customer Service Centre GmbH, Europaplatz 3, 69115 Heidelberg,
Germany. If you have any concerns regarding our products, please
contact ProductSafety@springernature.com

Printed and bound by CPI Group (UK) Ltd, Croydon, CR0 4YY
28/04/2026
02098539-0001